Lecture Notes in Mathematics

1632

Editors:
A. Dold, Heidelberg
F. Takens, Groningen

Springer

Berlin
Heidelberg
New York
Barcelona
Budapest
Hong Kong
London
Milan
Paris
Santa Clara
Singapore
Tokyo

Alexander Kushkuley   Zalman Balanov

Geometric Methods in Degree Theory for Equivariant Maps

Springer

Authors

Alexander Kushkuley
6 Carriage Drive
Acton, MA 01720, USA

Zalman Balanov
Department of Mathematics and Computer Science
Bar Ilan University
52900 Ramat-Gan, Israel

Cataloging-in-Publication Data applied for

Die Deutsche Bibliothek - CIP-Einheitsaufnahme

Kushkuley, Alexander:
Geometric methods in degree theory for equivariant maps /
Alexander Kushkuley ; Zalman Balanov. - Berlin ; Heidelberg ;
New York ; Barcelona ; Budapest ; Hong Kong ; London ;
Milan ; Paris ; Santa Clara ; Singapore ; Tokyo : Springer, 1996
  (Lecture notes in mathematics ; 1632)
  ISBN 3-540-61529-6
NE: Balanov, Zalman:; GT

Mathematics Subject Classification (1991):
Primary: 55M25, 55P91, 55S91, 58C30
Secondary: 55M35, 57Q91, 57R91, 57S15, 58E05, 58G45, 20C15

ISSN 0075-8434
ISBN 3-540-61529-6 Springer-Verlag Berlin Heidelberg New York

Typesetting: Camera-ready TeX output by the authors
SPIN: 10479811   46/3142-543210 - Printed on acid-free paper

To: Vita and Gabriel;
Larissa Tziporah, Sonya, Bronika and Jacob

Contents

0. Introduction

0.1. The mapping degree.

In these notes we are concerned with the mapping degree problem in the presence of group symmetries.

It seems to be a truism to say that the mapping degree is one of the most important topological tools employed in the study of nonlinear problems. The solvability of equations, multiplicity results, structure of solutions, bifurcation phenomena, geometric and (co)homological characteristics of functionals – this is a rather incomplete list of the subjects where the mapping degree plays a very important role.

The basic principles of the degree theory in the finite-dimensional case have been worked out by Kronecker, Poincaré, Brouwer and Hopf. Even nowadays the famous Hopf Classification Theorem and the Brouwer Fixed Point Theorem remain as brilliant examples of how the mapping degree works within topology as well as in its applications.

Significant contributions to the degree theory have been done by Borsuk and Leray-Schauder in the early thirties. Borsuk established that the degree of an odd map of a finite-dimensional sphere into itself is odd. By the same token, he observed for the first time that symmetries can lead to the restriction of possible values of the mapping degree. On the other hand, Leray and Schauder have extended the classical finite-dimensional degree theory to the infinite-dimensional case, defining it for maps of the form $I + A$, where I is the identity operator and A is a compact operator. This work was especially important from the viewpoint of extending the "application area" for the mapping degree methods.

Since the times of Borsuk, Leray and Schauder many mathematicians have been involved in developing degree theory. We refer the reader to [St] for an excellent survey of related results as well as an extensive list of references. Although there are many schemes reducing the study of nonlinear problems to calculating the mapping degree, computing (or even estimating) the degree in a practical way remains an actual problem in general. From the thirties until these days the degree problem for *equivariant* maps is attracting a good deal of attention.

0.2. The mapping degree and group symmetries.

Recall that if X and Y are metric spaces and G is a topological group acting on X and Y then a continuous map $f : X \to Y$ is said to be equivariant if $fgx = gfx$ for all $x \in X$ and $g \in G$ (the oddness presents the simplest example of equivariance with respect to $Z_2 = \{\pm I\}$).

Why should anyone be interested in the degree of equivariant maps? Apparently, there are at least two reasons for that. First, group symmetries appear in nonlinear problems in a very natural way. For instance, if an elliptic equation is defined on a domain $\Omega \subset R^n$ invariant with respect to some subgroup $G \subset O(n)$, then G acts naturally on the corresponding Sobolev space, and the integral operator, say, B, associated with the equation is G-equivariant. In addition, the eigenspaces of $B'(0)$ are G-invariant and (usually) are of finite dimensions, so that the standard Lyapunov-Schmidt procedure reduces the bifurcation problem to studying G-equivariant maps in finite-dimensional G-spaces. Note also that in the above case, calculating degrees of the equivariant maps is in close connection with studying geometric and (co)homological characteristics (like genus, G-category, cup-length, etc.) of the invariant functionals associated with the initial equation (see, for instance, [Bar1]).

Another source of "real life" symmetries comes from autonomous ordinary differential equations. If one looks for periodic solutions then the S^1-action on a space of periodic functions (induced by the time translation) should be taken into account (see, for instance, [IMV]).

A more "academic" example of how degrees of equivariant maps appear as an appropriate subject to study, comes from group representation theory. Namely, the problem of classifying representations of a compact Lie group G up to G-equivariant homotopy equivalence leads to equivariant versions of the Hopf Theorem (see, for instance, [Di1]).

Except for "external" motivations for studying degrees of equivariant maps (a wide field of applications) there are also "internal" ones connected with the following two observations. As is well-known, the difficulties in degree calculations increase with increasing dimension. From this point of view the presence of symmetries allows in many cases to decrease the dimension of the problem in question. For example, if a finite group G acts smoothly and semi-freely on smooth, compact, connected, oriented n-dimensional G-manifolds M and N then one can reduce the computation of the degree of an equivariant map $f : M \to N$ to studying the behavior of f on the set M^G of G-fixed points of M only. Namely, assume M^G and N^G are connected. If $\dim M^G \neq \dim N^G$ then $\deg f = 0 \pmod{|G|}$; if $\dim M^G = \dim N^G$ and, in addition, M^G, N^G are oriented then $\deg f \equiv \alpha \cdot \deg f|M^G \pmod{|G|}$, where α is relatively prime to $|G|$ and modulo $|G|$ is uniquely determined by the actions of G on M and N (cf. Chapter 3).

In addition, in many cases one can get an important information on the degree of an equivariant map only from algebraic characteristics of the corresponding actions. For instance, if a finite p-group G acts orthogonally without G-fixed points on a finite-dimensional sphere S then for any map $f : S \to S$ commuting with the G-action the following relation holds: $\deg f \equiv 1 \pmod{p}$.

During the past sixty years, the degree problem of equivariant maps has been attacked using various methods. After Borsuk, the following development of the theory was mostly due to P.A. Smith and M.A. Krasnoselskii. Smith introduced a special cohomology theory on a category of Z_p-spheres for a prime p which, in particular, was used in order to express degrees of equivariant maps via the homological characteristics of the corresponding actions (the so-called Smith indices), and via the degrees of the restrictions of the maps in question to the relevant sets of fixed points (if defined). This gives rise to the so-called "homological approach". Krasnoselskii discovered a deep connection between the "degree" problem for equivariant maps of Z_p-spheres (p-arbitrary) and the problem of equivariant extension of maps – essentially, the equivariant homotopy types appeared for the first time as an appropriate context for study (the so-called "geometric approach").

These notes are an attempt to describe in detail some recent achievements of the *geometric approach* and to present a comparative (albeit unavoidably incomplete) study of the results obtained by geometric and homological methods.

Since the literature on degree theory for equivariant maps is still growing enormously we only mention four books and one survey relevant to our discussion. These are:

- Dold's book [Do1] on the topology behind the finite-dimensional degree theory;

- Bredon's book [Bre] on the equivariant topology background;

- Ize, Massabó and Vignoli's book [IMV2] where the geometric approach for linear S^1-actions has been worked out in details (actually, equivariant maps of spheres of different dimensions are studied in [IMV2]. See also [IV], where linear actions of arbitrary abelian groups are considered, and [IMV1] for more general constructions);

- Borisovich and Fomenko's survey [BF] on homological methods in a degree theory for equivariant maps;

- Bartsch's book [Bar1] on the connection between the Borsuk type theorems and variational problems with symmetries:

Of course, we should mention Krasnoselskii's paper [Kr1] as the starting point for our research.

0.3. The geometric approach.

1. In the geometric approach, the degrees of two equivariant maps from a topological (compact, closed, connected, oriented) n-dimensional manifold M to the oriented n-dimensional sphere S are compared using equivariant extension theorems. More precisely, we can define a cylindric action of a group G on the cylinder $C = M \times [0,1]$ by setting it to be trivial on the segment $[0,1]$. We can also define a conic action of G on a ball B bounded by the sphere S (via the radial extension). Let O be the center of the ball. For equivariant maps $\Phi, \Psi : M \to S$ an equivariant map $f_0 : M \times \{0,1\} \to S \subset B$ is obviously defined. Let $F : C \to B$ be an equivariant extension of f_0 and $K = F^{-1}(O)$. If orientations on C and B are properly chosen then there exist fundamental classes $O_K \in H_n(C, C \backslash K)$ and $O_O \in H_n(B, B \backslash \{O\})$ which determine the degree of F (as a map of manifolds with boundaries) by the formula $F_*(O_K) = (\deg F)O_O$ (cf. e.g. [Dol], p. 268). Note also, that $\deg F = \pm(\deg \Phi - \deg \Psi)$. Let G be finite and let $(H_1), ..., (H_\ell)$ be all the orbit types of the G-action on M (and hence on C). Without loss of generality one can assume that every $g \in G$ either changes (simultaneously) orientations on M and S or preserves them. Suppose the extension F satisfies the following conditions:

(α) $K = \bigcup_{j=1}^{l} T_j$, $T_s \cap T_p = \emptyset$ if $s \neq p$;

(β) $T_j = G(K_j)$ for some compact K_j, $j = 1, ..., \ell$;

(γ) $H_j(K_j) = K_j$;

(δ) $g(K_j) \cap h(K_j) = \emptyset$ if $gh^{-1} \notin H_j$.

Now using (α) and the additivity of the degree, one gets $\deg F = \sum_{j=1}^{\ell} \deg F_j$ where F_j is the restriction of F on a sufficiently small neighbourhood of T_j. Bearing in mind that F is equivariant and using (β) $-$ (δ), we have $\deg F_j = a_j|G/H_j|$ where a_j is the degree of F in a small neighbourhood of K_j. From this we deduce the formula

$$(0.1) \qquad \deg F = \pm(\deg \Phi - \deg \Psi) = \sum_{j=1}^{\ell} a_j|G/H_j|\,,$$

and so

$$(0.2) \qquad \deg \Phi \equiv \deg \Psi \pmod{\mathrm{GCD}\{|G/H_j|_{j=1}^{\ell}\}},$$

which is the typical "comparison principle" result in the geometric approach.

If now one wants to estimate the degree of an arbitrary equivariant map $\Phi : M \to S$ it suffices to find only one equivariant map Ψ whose degree is easy to calculate, and to use formula (0.2). In many cases it is not hard to find the appropriate

Ψ. For instance, if M and S coincide as the G-spaces then one can take the identity map for Ψ. If $S^G = \{x \in S \mid gx = x \text{ for all } g \in G\} \neq \emptyset$ then one can set $\Psi(x) \equiv \text{pt} \in S^G$.

This approach has been first realized by M.A. Krasnoselskii [Kr1] for the case when $G = Z_p$ acts freely on a sphere.

2. The above discussion gives rise to the following problem. Let G be a compact Lie group, let X, Y be a couple of metric G-spaces, and let $A \subset X$ be a closed invariant subspace. What are conditions on X and Y which imply that an equivariant map $f : A \to Y$ has an equivariant extension over X? This problem was addressed by several authors, e.g. J. Jaworowski [Ja1, Ja2], R. Lashof [La], M. Madirimov [Mad1, Mad2] and others. All these authors used the reduction of the above problem to the problem of extending sections of fiber bundles associated with the maps in question. In this book we develop another general approach which is intuitively easier and allows us to obtain stronger extension results in certain cases. What is more important, this extension technique provides some means for controlling the extension map in a manner required by the "comparison principles" like the one described above. Again, the idea behind this approach can be traced back to the original paper of M.A. Krasnoselskii [Kr1].

The key to the extension results we are looking for is the following

Definition. Let a topological group H act on a metric space E. Let $D_0 \subset E$ be open in its closure D. Then D is said to be a *quasi-fundamental domain* of the H-action on E if the following conditions are satisfied:

(a) $H(D) = E$;

(b) $g(D_0) \cap h(D_0) = \emptyset$ $(g \neq h; \ g, h \in H)$;

(c) $E \backslash H(D_0) = H(D \backslash D_0)$.

If E is finite-dimensional and the following additional condition holds

(d) $\dim D = \dim E/H$; $\dim(D \backslash D_0) < \dim D$; $\dim H(D \backslash D_0) < \dim E$

then D will be called a *fundamental domain* for the H-action on E.

Note that, if H is a discrete group then one can set D_0 to be the interior of D. Hence the definition above naturally complies with the classical one (cf. e.g. [DFN], p. 169).

It turns out that a (quasi-)fundamental domain exists for any free action of a compact Lie group on a metric space.

Assume now that a compact Lie group G acts on a metric space X and let $A \subset X$ be a closed invariant subset such that the action of G on $X \backslash A$ is free. By the above observation there exists a (quasi-)fundamental domain $D^{(0)}$ of the G-space $X \backslash A$.

Let $D_0^{(0)}$ be the corresponding open subset of $D^{(0)}$ and let $X_1 = A \cup G(D^{(0)} \setminus D_0^{(0)})$. Applying the above observation once again to $X_1 \setminus A$ we get X_2, etc. So one has a closed invariant filtration $X = X_0 \supset X_1 \supset X_2 \ldots \supset A$. If $X \setminus A$ is finite-dimensional then this filtration is finite. Let Y be another metric G-space. It turns out that if $X \setminus A$ is finite dimensional then any equivariant map $A \to Y$ extends over X if for all $i = 1, 2, \ldots$ any equivariant map $X_i \to Y$ has a (non-equivariant) extension over $X_i \cup D^{(i-1)}$. The same is true for extensions of equivariant homotopies.

Combining the last argument with the standard induction over the orbit types (see, for instance, [Di1]) leads to a rather general equivariant version of the well-known Kuratowski-Dugundji Extension Theorem. In particular, if for any stationary subgroup H of the action of G on M one has dim $\{x \in M \mid hx = x$ for any $h \in H\} \leq n(H)$ and if the set $S^H = \{y \in S \mid hy = y$ for any $h \in H\}$ is locally and globally k-connected for each $k = 0, 1, 2, \ldots, n(H) - 1$, then the existence of an equivariant extension with properties (α) - (δ) required by the comparison principle follows immediately from the considerations above.

Using this scheme we strengthen the corresponding degree results by Krasnosel-skii [Kr1], Zabrejko [Za1, Za2], Bowczyc [Bow1, Bow2], Dold [Do2], Daccach [Dac] and others.

To some extent, this approach can be characterized as "geometric equivariant obstruction theory without CW-complexes".

3. The next step of our program is to improve the general geometric approach in such a way that one could treat the following problems:

(a) to replace in formulas (0.1) and (0.2), a finite group (lengths of the orbits) by an arbitrary compact Lie group (Euler charactaristics of the orbits);

(b) to eliminate the connectedness conditions with respect to the sets S^H;

(c) to express explicitly the degree of an equivariant map via geometric characteristics of actions and degrees (if defined) of the restrictions of the given map to appropriate fixed point sets.

To this end, assuming G to be an arbitrary (not necessarily finite) compact Lie group, we impose the following additional conditions: M is a smooth G-manifold and S is a G-representation sphere. These assumptions allow us to take advantage of some standard (but important) tools from Riemannian G-geometry (invariant tubular neighborhood, normal slice, invariant foliations, etc.), algebraic topology (cap-product, Thom class, etc.) and piecewise linear topology. The main idea remains the same: to construct an equivariant extension F in such a way that the set $F^{-1}(0)$ is "computable". But now we provide F with more "delicate" properties than those formulated in $(\alpha) - (\delta)$ (for the precise formulation see Lemma 3.8).

Our approach is essentially based on the following three observations.

1) It is well-known (see, for instance, [Bre]) that if M is a compact smooth G-manifold and $N \subset M$ is a G-submanifold, then there exists an invariant tubular neighborhood of N in M. In particular, this means that there exists an invariant one-dimensional foliation around N. If now N is the union of all non-principal orbits for the action of G on M, then N is not a submanifold of M in general, so that it may happen that there does not exist a tubular neighborhood around N. It turns out, however, that there exists an invariant one-dimensional foliation around N in this case as well.

2) Let U be an oriented n-dimensional manifold, V an n-dimensional vector space and D its k-dimensional subspace. Assume W_1 is an open subset of V such that W_1 is contratible to $D \setminus 0$ and $W_1 \bigcup (V \setminus D) = V \setminus 0$. Let $f : U \to V$ be a continuous map such that $K = f^{-1}(0)$ is compact. Suppose, finally, that there exists an open subset $U_1 \subset f^{-1}(W_1)$ such that $U_1 \bigcup (U \setminus f^{-1}(D)) = U \setminus K$. Denote by τ_D^V the Thom class of D in V.

It turns out that $\deg_0 f = 0$ if $f^*(\tau_D^V) = 0$. In particular, under the above conditions $\deg_0 f = 0$ provided $H^{n-k}(U, U \setminus f^{-1}(D)) = 0$.

3) The last observation is concerned with "general position" theorems in the equivariant context. Let V be an orthogonal $(d+1)$-dimensional representation of a finite group G and B^{d+1} the unit ball in V. Let G act freely on a compact $(d-k)$-dimensional manifold X ($k \geq 1$). For any finite set of linear subspaces $L_j \subset V$, $j = 1, \ldots, m$, there exists an equivariant map f from X to B^{d+1} such that $\dim f^{-1}(G(B^{d+1} \bigcap L_j)) \leq \dim L_j - k - 1$ for all $j = 1, 2, \ldots, m$, provided $\dim L_j \geq k$.

These three observations in compliance with the above mentioned equivariant extension technique based on the notion of fundamental domain lead to an essential strengthening of the comparison principle in directions (a) - (c). In particular, we generalize the corresponding results by Nirenberg [Ni2], Marzantowicz [Mar1], Wei Yue-Ding [We], Dancer [Dan], Lück [Lü], Komiya [Kom], Fadell, Husseini and Rabinowitz [FHR] and others; we also strengthen in certain cases the results by Ize, Massabo and Vignoli [IVM2, IV]; finally, we clarify the geometric nature of the results by Borisovich, Izrailevich and Fomenko (Schelokova) [Sc4, BF].

For the precise formulations of our results we refer the reader to Section 3.1. Below we present two corollaries which can be stated without additional explanations.

Let $G = T$ be a torus and let $\Phi, \Psi : M \to S$ be T-equivariant maps. Let $M_1, M_2, \ldots, M_m$ be the connected components of M^T.

(a) For each j, $\dim M_j = \dim S^T$, there exists an integer $\alpha_j = \alpha(M_j, S^n)$

completely defined by the G-actions on M and N such that

$$\deg \Phi - \deg \Psi = \sum_j \alpha_j \cdot (\deg \Phi | M_j - \deg \Psi | M_j);$$

(b) if $\dim M_i \neq \dim S^T$ for all i then $\deg \Phi = \deg \Psi$ is uniquely determined by the G-action on M and S.

Assume now a finite p-group G acts smoothly on compact, connected, oriented n-dimensiomal manifolds M and N. Suppose that $N^G \neq \emptyset$ and all the fixed point sets N^H, $(H) \in \mathrm{Or}(N)$, are connected and oriented. Let $\{M_i | i = 1, 2, ..., m\}$ be the set of connected componenets of M^G with $\dim M_i = \dim N^G$. Then for any equivariant map $f : M \to N$,

$$\deg f \equiv \sum_i \alpha_i \deg(f | M_i) \pmod{p},$$

where the numbers α_i are uniquely determined modulo p. In particular, if $m = 0$ then $\deg f \equiv 0 \pmod{p}$.

It should be noticed that we are interested in equivariant maps of G-manifolds of the same dimension. Therefore, we are dealing with the Brouwer degrees only. At the same time, the authors of [IMV1, IMV2, IV] deal with equivariant maps of G-representation spheres of different dimensions, and calculate the so-called "equivariant degree" which coincides with the Brouwer degree if dimensions of the spheres coincide. From this point of view certain results obtained in [IVM1, IVM2, IV] are, of course, more general than those presented in our monograph. However, it is easy to see that one can use the methods developed in our monograpgh to study the above mentioned situation as well.

4. One of the natural applications of the stream of ideas discussed above is the so-called Equivariant Hopf Theorem.

Recall a classical theorem of H. Hopf (see, for instance, [Di2], p. 122). Let M be a closed, compact, connected, oriented n-dimensional manifold and S an oriented n-dimensional sphere. Hopf's theorem asserts that two continuous maps from M to S are homotopic iff their degrees are equal, and, in addition, that any integer can be realized as the degree of some map from M to S. Suppose now that a compact Lie group G acts on M and S. Classification of equivariant maps $M \to S$ up to equivariant homotopy can not be achieved in the same straightforward way as in the non-equivariant case. As an example, suppose that G is a finite group acting orthogonally on vector spaces V and W. Denote by $S(V)$ and $S(W)$ the correponding representation spheres. Suppose that for all subgroups $H \subset G$

the dimensions of fixed point sets V^H and W^H are equal. Consider the following statement :

(*) *G-equivariant maps* $f_1, f_2 : S(V) \to S(W)$ *are equivariantly homotopic if and only if*

$$\deg(f_1|S(V)^H) = \deg(f_2|S(W)^H)$$

for all subgroups H of G.

Although this statement is not true in general (see [Ru]), there exists a rather general set of conditions on G-spaces V and W which ensure its validity. These conditions can be obtained as a corollary of the so-called Equivariant Hopf Theorem presented by tom Dieck in [Di1,Di2] and generalized by Tornehave [To] and Laitinen [Lai]. An equivariant cohomology theory has been used as the main tool in [Di1, Di2, Lai, To]. In these notes we discuss a more straightforward approach to the Equivariant Hopf Theorem based on combining the usual (non-equivariant) obstruction theory with the fundamental domain technique. In particular, this enables us to obtain *necessary and sufficient* conditions for statement (*) and, in addition, to strengthen the results on equivariant homotopy classification obtained in [Di1,Di2,To,Lai].

5. Our final remark is concerned with the infinite dimensional aspect of the degree problem for equivariant maps.

In accordance with the classical approach by Leray and Schauder, to carry out the finite-dimensional results to completely continuous vector fields in Banach spaces one should solve the following problem. Let $\Phi = I + A$ be a completely continuous vector field defined on the closure of a bounded region Ω in a Banach space E. Let G be a compact Lie group. Assume Φ is equivariant with respect to a pair of linear representations of G in E. Given $\varepsilon > 0$ one should construct a finite dimensional operator $A_n : \overline{\Omega} \to E$ such that:

1) A_n is equivariant;

2) $||A - A_n|| < \varepsilon$.

This is not a difficult problem if one deals with vector fields which are equivariant with respect to one representation only. However, in the case of two representations the "co-existence" of equivariance with infinite dimension leads to a "conflict". Namely, even in the case when G is a cyclic group it may happen that given a finite-dimensional subspace $E^k \subset E$ there is no finite-dimensional subspace $E^d \supset E^k$, which is invariant with respect to the pair of G-representations *simultaneously*.

In these notes we develop a method, based on combining the classical Leray-Schauder technique with some ideas from the theory of gaps between linear subspaces, which allows us, in certain cases, to overcome this conflict.

0.4. Overview.

The book consists of five chapters.

The first chapter is devoted to studying the equivariant extension problem. In the first section we present auxiliary information from equivariant topology. In the second section we prove the existence of fundamental domains in a rather general situation. By means of this result we prove the Equivarint Kuratowski-Dugundji Theorem in the third section.

In the second chapter assuming M to be a closed, compact, connected, oriented, topological n-dimensional manifold, and S to be an oriented n-dimensional sphere we study degrees of maps $M \to S$ equivariant with respect to *topological* actions on M and S. The first section is devoted to the general comparison formula for degrees of equivariant maps (G is a finite group). Some special cases and generalizations of this formula (p-group actions, free actions, torus actions, etc.) are considered in the second section. We conclude the chapter with some counterexamples which show that our hypotheses are sharp in some respect.

In the third chapter we assume that G is an arbitrary compact Lie group, M is a smooth (closed, compact, connected, oriented) G-manifold and S is a G-representation sphere. Under these assumptions we get sharper results than those stated in Chapter 2. In particular, we remove the connectedness conditions with respect to the G-action of on S, and in many cases give precise restrictions on the possible values of degrees of equivariant maps from M to S.

This chapter is organized as follows. In the first section using our results from Chapters 1 and 2 we introduce some integer-valued characteristics connected with the actions of G on M and S. One may consider these characteristics as the geometric analogs of the so-called equivariance indices introduced by T. Fomenko (Schelokova) in [Sc4] (see also [BF], [Di1], [Di2]). In terms of these characteristics we formulate our main results and present some corollaries for p-group actions, torus actions, semi-free actions, abelian group actions, etc. Taking an arbitrary smooth manifold N instead of S and assuming a group G is acting on N so that $N^G \neq \emptyset$, we use some straightforward arguments in order to show that most of our results remain valid in this situation.

The second section is auxiliary. Here we present some properties of the cap-product and Thom class allowing us to deal with "bad" orbit types in M (those for which $\dim M^K > \dim S^K$).

In the third section we present the above mentioned invariant foliation and equivariant general position lemmas. These lemmas together with the "elimination" technique based on the usage of the Thom class are main ingredients in our approach. They come together in the forth section where we give a proof of one of our main degree results assuming G to be a finite group.

In the fifth section we extend our result to the case of an arbitrary compact Lie group actions. In the sixth section we consider equivariant maps from one G-manifold to another G-manifold without assuming the second manifold to be a sphere. As a particular case, we consider here abelian group actions.

The fourth chapter is devoted to the degree problem for completely continuous vector fields in Banach spaces. In the first section we develop some machinery for solving the conflict between equivariance and infinite dimensionality. In the second section we use this technique to get our degree results.

In the last chapter we present some applications of the methods developed in the previous chapters.

In the first section we consider a semi-linear elliptic boundary value problem which is associated with the corresponding linear problem of positive Fredholm index. Under some symmetry assumptions we prove the existence of solutions of arbitrarily large norm in the corresponding Hölder space. We follow the scheme by P. Rabinowitz [Ra1] (see also [Mar1]).

In the second section we give a lower estimate for the genus of the free part of a finite-dimensional sphere S with a compact Lie group action. To treat this problem we modify the well-known geometric aproach by M. Krasnoselskii [Kr1, KZ]. We apply the obtained result to the irreducible $SO(n)$-representations in spherical harmonics.

In the third section we present an equivariant version of the Hopf Theorem on the homotopy classification of mappings from a manifold to a sphere. Some illustrative examples are considered.

The fourth section is devoted to the Borsuk-Ulam type theorems on the nonexistence of an equivariant mapping from an n-dimensional free G-sphere to an m-dimensional one if $n > m$; we consider a situation of non-free actions on manifolds.

In the fifth section we give an elementary proof of the well-known theorem of Atiyah-Tall [AT]:

Let V and W be two finite-dimensional orthogonal representations of a p-group G ($\dim V = \dim W$), and let $S(V)$ and $S(W)$ be the unit spheres in V and W respectively. Then there exists a G-equivariant map $f : S(V) \to S(W)$ with

$\deg f \neq 0$ (mod p) iff irreducible components of V and W are conjugate in pairs by elements (possibly different) of the corresponding Galois group.

Certain questions concerning G-equivariant maps of G-manifolds related to this theorem are also discussed.

0.5. Acknowledgements.

The authors are grateful to S. Antonian, T. Bartsch, Y. Bregman, M. Goresky, M. Krasnoselskii, R. Palais, K. Pawałowski, M. Postnikov, D. Puppe, A. Shostak, Y. Smirnov, H. Steinlein and C. Terng for helpful discussions.

We are especially grateful to S. Brodsky and P. Zabrejko for long stimulating conversations, and to A. Dold, J. Ize and Ju. Rudyak who have improved our understanding of the subject.

As usual, only the authors are responsible for any possible omissions and/or errors.

The work on these notes started during the visit of the second author to Brandeis University in 1991. We are grateful to the members of the Mathematics Department for their hospitality and to R. Palais for his invitation. The support of the University of Tel Aviv was essential for this visit.

The second author acknowledges the support of the Israel Ministry of Absorbtion and the Minerva Foundation in Germany through the Emmy Noether Institute at the Bar-Ilan University.

These notes were finished during the visits of the second author to the University of Heidelberg and to the University of Munich as an Alexander von Humboldt Fellow. The second author is thankful to the AvH Foundation for its support, to A. Dold and H. Steinlein for their invitations and to the members of the Mathematics Institutes for their hospitality.

The authors are thankful to H. Steinlein for assistance in preparing the manuscript for publication.

Chapter 1

FUNDAMENTAL DOMAINS AND
EXTENSION OF EQUIVARIANT MAPS

In accordance with the general scheme outlined above this chapter is
devoted to the problem of the extension of equivariant maps. Nevertheless,
some results may be interesting on their own.

In the first section we present some auxiliary material from G-space
theory and topology. The main tool of this chapter is introduced in the
second section. Namely, we define a notion of a fundamental domain of a
free action of a topological group on an arbitrary metric space and prove
its existence for compact Lie groups. In the third section we describe a
geometric approach to the problem of the extension of equivariant maps
which is based on the notion of a fundamental domain. As an illustration
of our technique we give a proof of the general version of the equivariant
Kuratowski-Dugundji Theorem.

1.1. Auxiliary information

1.1.1. A transformation group is a triple (G, K, μ) where G is a topological group,
K is a Hausdorff topological space and $\mu : G \times K \to K$ is a continuous map such
that:

1) $\mu(g, \mu(h, x)) = \mu(gh, x)$ for all $g, h \in G$ and $x \in K$;
2) $\mu(e, x) = x$ for all $x \in K$, where e is the identity of G

(see, for instance, [Bre], p. 32; [Di2], p. 2). The map μ is called an action of G on K.
The space K together with a given action μ of G is called a G-space. We shall often
use the same notations for a G-space as for the underlying Hausdorff topological
space, regarding μ as understood. For the simplification of the notations we shall
often write gx instead of $\mu(g, x)$. For $H \subset G$ and $N \subset K$ we set $H(N) = \{gx :
g \in H, x \in N\}$. A set N is called invariant with respect to the action of G (or
G-invariant) if $G(N) = N$.

For every $x \in K$ we denote by G_x a stationary subgroup of the point x, i.e. a
set $\{g \in G : gx = x\}$. A family of all stationary subgroups of the action of G on

K is denoted by Iso(K). The orbit type (H) of G-space K is the set of conjugacy classes of some stationary subgroup H.

An action of G on K is called free, if G_x is trivial for all $x \in K$, and semi-free, if G_x is trivial or coincides with G $(x \in K)$. Free and semi-free actions give us the most simple examples of the actions with a finite number of orbit types. Note that by the famous result of Mann [Man] every action of a compact Lie group on an oriented manifold with finitely generated homology groups has a finite number of orbit types.

Let X be a G-space and let $x \in X$. A subspace $G(x) = \{gx \in X : g \in G\}$ is called an orbit of the point x (with respect to the action of group G). It is easy to see that if H is a stationary subgroup of $x \in X$ then the orbit of x is homeomorphic to G/H and orbits of the two points $x, y \in X$ are homeomorphic iff their stationary subgroups are conjugate (see, for instance, [Bre], pp. 37, 40). Therefore, sometimes when we speak about an arbitrary orbit of type (G_x) we shall use the notation G/G_x for $G(x)$.

Let us denote by X/G the set of the orbits of a G-action on X and let $p : X \to X/G$ be a natural orbit map defined by the formula $p(x) = G(x)$. Then p defines a quotient topology on X/G in a standard way. It is well-known (see, for instance, [Bre], p. 38) that if G is a compact group then $p : X \to X/G$ is a closed map.

The action of G on K is called effective if $\{g \in G : gx = x$ for all $x \in K\} = \{e\}$.

If H is a subgroup of G we shall set $K^H = \{x \in K : hx = x, \ h \in H\}$, $K_H = \{x \in K : G_x = H\}$, $K^{>H} = K^H \backslash K_H$ and $K_{(H)} = \{x \in K : (G_x) = (H)\}$.

Let Γ be a group, H be a subgroup of Γ and $N(H)$ be a normalizer of H in Γ. The quotient $W(H) = N(H)/H$ is called the Weyl group of H. Clearly, the action of $W(H)$ on K^H is correctly defined and this action is free on K_H (see, for instance, [Bre], pp. 45, 90).

As usual, if G is a compact Lie group then $K \subset G$ is said to be a torus if K is a closed connected abelian subgroup. If K is a torus and there is no a torus $K' \supset K$, then K is said to be a maximal torus of G. We will denote by T a maximal torus of G; the integer $\dim T$ is said to be a rank of G (denoted by rk G). A closed subgroup $H \subset G$ is called a subgroup of maximal rank if it contains a maximal torus. By definition rk G coincides with the rank of the connected component of the unity of G. In particular, if G is a finite group then the maximal torus of G coincides with the unity of G, and all subgroups of G are of maximal rank. As is well-known, for any subgroup $H \subset G$ of maximal rank (for instance, $H = T$) the group $W(H)$ is finite [Bo2].

The following result is well-known as Gleason's Lemma.

Lemma 1.1 (see [Bre], p. 88). *Let G be a compact Lie group and X be a metric G-space with one orbit type (H). Then the orbit map $p : X \to X/G$ is the projection in a fiber bundle with fiber G/H.*

Below we will need an obvious

Lemma 1.2. *Suppose that $p : T \to B$ is a fiber bundle. Suppose, further, that B is a disjoint union of its open subsets $\{U_\alpha\}_{\alpha \in I}$ such that p is trivial over U_α for all $\alpha \in I$. Then p is trivial over B.*

Let Y be another G-space. A continuous map $\varphi : K \to Y$ is said to be equivariant (G-equivariant) if $\varphi g x = g \varphi x$ for all $x \in K$ and $g \in G$. An equivariant map $\varphi : K \to Y$ which is also a homeomorphism is called an equivalence of G-spaces. It is clear that if $\varphi : K \to Y$ is an equivariant map then $G_x \subset G_{\varphi(x)}$.

1.1.2. Here we recall some standard notions from retract theory.

Let X be a metric space and $Y \subset X$. Then a continuous map $f : X \to Y$ is said to be a retraction if $f(x) = x$ for all points $x \in Y$; the subset Y is called a retract of X.

A closed subset X_0 of the metric space X is called a neighbourhood retract in X if X_0 is a retract of some open subset of X containing X_0.

A metric space M is called an AR-space if for each homeomorphism h mapping M onto closed subset $h(M)$ of a metric space M_1, the set $h(M)$ is a retract of M_1. A metric space M is said to be an ANR-space if for each homeomorphism h mapping M onto closed subset $h(M)$ of a metric space M_1, $h(M)$ is a neighbourhood retract in M_1.

It is well-known ([Bor1], p. 77) that Y is an AR-space (ANR-space) iff for every closed subspace T of the metric space T_1 and any continuous map $f : T \to Y$ there exists a continuous extension $f_1 : T_1 \to Y$ of f (a continuous extension $f_2 : U \to Y$ of f over some neighbourhood U of T in T_1).

A set $B \subset M$ is said to be contractible with respect to the metric space M if an imbedding $i : B \to M$ is homotopic to the constant map $j : M \to x_0 \in M$; if, in addition, $M = B$ then B is called contractible. The space B is said to be locally contractible if for every point $x_0 \in B$ and any of its neighbourhood $U \subset B$ there exists a neighbourhood $V \subset U$ ($V \ni x_0$) which is contractible with respect to U.

It is well-known that every ANR-space is locally contractible ([Bor1], p. 87) and that a space is an AR-space if it is a contractible ANR-space ([Bor1], p. 96).

A space B is called k-connected ($k = 0, 1, 2, \ldots$) if every continuous map $f : S^k \to B$ is homotopic to the constant map, where S^k is a k-dimensional sphere. A space B is called locally k-connected ($k = 0, 1, 2, \ldots$) if for any point $x_0 \in B$ and its neighbourhood $U \subset B$ there exists a neighbourhood $V \subset U$ ($x_0 \in V$) such that any continuous map $f : S^k \to V$ is homotopic (in the space U^{S^k}) to the constant map. It is easy to see ([Bor1], p. 30) that any contractible (locally contractible) space is k-connected (locally k-connected) for every $k = 0, 1, 2, \ldots$

A metric space Y is called an ENR-space if an ANR-space $X \subset R^n$ exists which is homeomorphic to Y. It is well-known that locally compact separable ANR-spaces of finite dimension are ENR-spaces (see [Do1], p. 84). Let X be a metric G-space and $Y \subset X$ be its closed G-subspace. Then Y is said to be a G-retract of X if there exists a G-equivariant retraction $f : X \to Y$. A G-space X is said to be a $G - ENR$-space if it is equivalent to a G-retract of some open G-subset in some Euclidean G-space with an orthogonal action of G. For instance, a differentiable G-manifold with a finite number of orbit types is a $G - ENR$-space ([Di1], p. 86).

1.1.3.   Here we shall collect some facts from dimension theory. Always below we keep in mind a covering dimension (see, for instance [En], p. 266).

Lemma 1.3 (see [En], Theorem 7.3.11). *If K is a metric space and $\dim K \leq n$ then for any closed $F \subset K$ and any open $V \subset K$ ($F \subset V$) there exists an open set $U \subset K$ such that $F \subset U \subset \overline{U} \subset V$ and $\dim \partial U < n$.*

Lemma 1.4 (see [En], Theorem 7.2.3). *Let $\{F_\alpha\}_{\alpha \in I}$ be a locally finite covering of the metric space K such that for all $\alpha \in I$ the sets F_α are closed and $\dim F_\alpha \leq n$. Then $\dim K \leq n$.*

Suppose that X is a finite-dimensional G-space and G is a compact Lie group. Using the definition of covering dimension and the Morita Theorem (see [Mor]), which states that $\dim(K \times [0,1]) = \dim K + 1$ for any metric space K, one can easily prove

Lemma 1.5. *If X is a free G-space then $\dim(X/G) = \dim X - \dim G$.*

Lemma 1.6. *Suppose that a compact Lie group G acts on a finite-dimensional metric space X with a finite number of the orbit types, F is an invariant compact subset of F and W is an invariant neighbourhood of F. Then there exists an invariant neighbourhood $U_F \supset F$ such that $U_F \subset W$ and $\dim \partial U_F < \dim X$.*

This lemma is a trivial consequence of Lemmas 1.1 and 1.4 and we omit the proof.

We'll finish this subsection with properties of locally finite families which will be used below.

Lemma 1.7 (see [En], Theorem 1.5.18). *Suppose that $\{U_\alpha\}_{\alpha\in I}$ is a locally finite open covering of a metric space. Then there exists an open covering $\{V_\alpha\}_{\alpha\in I}$ such that $\overline{V}_\alpha \subset U_\alpha$ for all $\alpha \in I$.*

Lemma 1.8 (see [En], p. 46). *Let $\{F_\alpha\}_{\alpha\in I}$ be a locally finite family of subsets of a metric space. Then*

$$\partial\left(\bigcup_{\alpha\in I} F_\alpha\right) \subset \bigcup_{\alpha\in I}(\partial F_\alpha) \ .$$

1.1.4.   Here we'll recall the definition and standard properties of the degree of continuous maps of manifolds with boundary. All the (co)homology groups are considered over Z (unless otherwise stipulated).

Let M be an oriented n-dimensional manifold with boundary ∂M,

$$O : M\backslash\partial M \to \bigcup_{P\in M\backslash\partial M} H_n(M, M\backslash P)$$

be some orientation on it ([Dol], p. 264) and K be a compact subset in $M\backslash\partial M$. Then there exists ([Dol], p. 260) a unique element $O_K \in H_n(M, M\backslash K)$ satisfying the condition: the inclusion homomorphism $i_*^P : H_n(M, M\backslash K) \to H_n(M, M\backslash P)$ takes O_K into $O(P)$, for every $P \in K$. This element O_K is called the fundamental class around K. In particular, if M is a compact manifold without boundary then there exists a fundamental class $O_M \in H_n(M, M\backslash M) = H_n(M, \emptyset) = H_n(M)$, and if K is connected and non-empty, then $H_n(M, M\backslash K) = Z$ and O_K is a generator of $H_n(M, M\backslash K)$.

Definition. *Let $f : (M_1, \partial M_1) \to (M_2, \partial M_2)$ be a continuous map between oriented n-dimensional manifolds with boundary, such that $f(\partial M_1) \subset \partial M_2$. Suppose, further, that K is a connected compact non-empty subset in $M_2\backslash\partial M_2$ such that $f^{-1}(K)$ is compact. Then ([Dol], p. 268) $f_* : H_n(M_1, M_1\backslash f^{-1}(K)) \to H_n(M, M\backslash K)$ takes the fundamental class $O_{f^{-1}(K)}$ into an integer multiple of O_K; this integer is called the degree of f over K, and denoted by $\deg_K f$. Thus, $f_*(O_{f^{-1}(K)}) = (\deg_K f)\cdot O_K$. If $M_2\backslash\partial M_2$ is connected then ([Dol], p. 268) the number $\deg_K f$ is the same for all non-empty connected compact sets $K \subset M_2\backslash\partial M_2$. It is called the degree of f, and is denoted $\deg f$.*

For certain applications it is sometimes more convenient to consider maps

$$f : (M, \partial M) \to (R^{n+1}, R^{n+1}\backslash\{0\}) \ .$$

Since homology groups of $R^{n+1}\backslash\{0\}$ coincide with the ones of an n-dimensional sphere the degree $\deg_0 f$ is correctly defined (see, for instance, [KZ], [Ni1]).

Let $(M_1, \partial M_1)$ and $(M_2, \partial M_2)$ be a pair of compact oriented n-dimensional manifolds with boundary and the orientations of the boundaries are induced by the orientations of the manifolds (see, for instance, [Do1], p. 257). If $f : (M_1, \partial M_1) \to (M_2, \partial M_2)$ is a continuous map, $f(\partial M_1) \subset \partial M_2$ and M_2 and ∂M_2 are connected then the degree of $f|\partial M_1$ is correctly defined.

Lemma 1.9 ([Do1], p. 257, 267; [Sp], p. 301). *Under the above assumptions* $\deg f = \deg f|\partial M_1$.

The following properties of degree are well-known.

Lemma 1.10 ([Do1], p. 269). *Let* $f : (M_1, \partial M_1) \to (M_2, \partial M_2)$ *be as above, and let* K *be a connected non-empty compact set in* $M_2 \backslash \partial M_2$. *In addition, let* $M_1 \backslash \partial M_1$ *be a finite union of open sets* $\{M_1^\alpha\}$, $\alpha = 1, \ldots, r$ *such that the sets* $K_1^\alpha = f^{-1}(K) \cap M_1^\alpha$ *are mutually disjoint. Then*

$$\deg_K f = \sum_\alpha \deg_K f|M_1^\alpha .$$

Lemma 1.11 ([Do1], p. 268). *Let* M_1 *and* M_2 *be as above and let* $f : M_1 \to M_2$ *be a homeomorphism. Then* $\deg_K f = \pm 1$ *(according to these two cases* f *is called orientation – preserving or – reserving.*

Lemma 1.12 ([Do1], p. 268). *Suppose that* M_1, M_2 *and* M_3 *are connected compact oriented* n-dimensional manifolds. *Suppose, further, that continuous maps* $f : M_1 \to M_2$ *and* $h : M_2 \to M_3$ *are given.*

Then $\deg(h \circ f) = \deg h \cdot \deg f$.

Remark 1.1.

(a) Let M_1, M_2 be as in Lemma 1.12. Assume a finite group G acts on M_1 and M_2. By Lemma 1.11 these actions define the so-called orientation homomorphisms $e_{G,M_1} : G \to Z^* = \{1, -1\}$ and $e_{G,M_2} : G \to Z^* = \{1, -1\}$ (see also subsection 5.3.2 where the general case of CW-complexes is considered). The actions of G on M_1 and M_2 are said to be concordant if $e_{G,M_1} = e_{G,M_2}$. Let $f : (M_1, \partial M1) \to (M_2, \partial M_2)$ be an equivariant map. If the actions are not concordant then $\deg f = 0$. Indeed, by Lemmas 1.11 and 1.12 for any $g \in G$ we must have:

$$\deg f = e_{G,M_1}(g^{-1})(\deg f)e_{G,M_2}(g),$$

and if $e_{G,M_1}(h) \neq e_{G,M_2}(h)$ for some $h \in G$ then $\deg f = -\deg f$. Hence, when we speak about the degree of an equivariant map we assume that the actions are concordant.

(b) Let M_1 and M_2 be as above. Let $f : M_1 \to M_2$ be a map such that the set $f^{-1}(p) = \{q_1, q_2, ..., q_m\}$ is finite. Let U_i, $i = 1, 2, ..., m$, be open neighborhoods of points q_i, $i = 1, 2, ..., m$, such that $U_i \cap U_j = \emptyset$ if $i \neq j$. Then by Lemma 1.10

$$\deg f = \sum_1^m \deg f | U_i.$$

Now suppose a finite group G acts on M_1 and M_2 in such a way that $p \in M_2^G$ and $f^{-1}(p) = G(q)$ is a G-orbit (assume that $q = q_1$). Then

$$\deg f = |G/G_q| \cdot \deg(f|U_1).$$

Indeed, taking into account Remark 1.1(a) and Lemmas 1.10 - 1.12 note that for any $h \in G$ such that $q_i = h(q)$

$$\deg f | U_i = e_{G,M_1}(h^{-1})(\deg f|U_1)e_{G,M_2}(h) = \deg(f|U_1).$$

Remark 1.2. Let X and Y be topological spaces. Then the join $X * Y$ is the space obtained from $X \times Y \times [0, 1]$ by identifying $\{x\} \times Y \times \{0\}$ to a point for each $x \in X$, and identifying $X \times \{y\} \times \{1\}$ to a point for each $y \in Y$.

As is well-known, if S^n and S^m are n- and m-dimensional spheres (n, $m \geq 0$) then $S^n * S^m$ is homeomorphic to a sphere S^{n+m+1}. Further, for any $z \in S^{n+m+1} \subset R^{n+m+2} = R^{n+1} \times R^{m+1}$ and $t \in [0, 1]$ one has the following representation:

$$z = \cos(\pi t/2) \cdot x + \sin(\pi t/2) \cdot y$$

with $x \in S^n$ and $y \in S^m$. Given $f : S^n \to S^n$ and $g : S^m \to S^m$ define their join $f * g : S^{n+m+1} \to S^{n+m+1}$ by

$$(f * g)(z) = \cos(\pi t/2) \cdot f(x) + \sin(\pi t/2) \cdot g(x)$$

(see [Do1], p. 66). It is easy to see that $f * g = (f * I) \circ (I * g)$, $\deg(f * I) = \deg f$ and $\deg(I * g) = \deg g$ (here I stands for the identity map). Now from Lemma 1.12 it follows immediately

$$\deg(f * g) = \deg f \cdot \deg g.$$

Note that if X and Y are G-spaces, then so is $X * Y$ via

$$g(x, y, t) = (g(x), g(y), t).$$

It is easy to see that if G acts orthogonally on S^n and S^m then the action of G on $S^n * S^m$ is equivalent to the orthogonal action of G on S^{n+m+1}. Moreover, if under the above assumptions $f : S^n \to S^n$ and $g : S^m \to S^m$ are equivariant maps then so is $f * g : S^n * S^m \to S^n * S^m$.

Lemma 1.13 ([Dol], p. 271). *If M_1, M_2 are compact oriented n- and m-dimensional manifolds (respectively) and $I \subset M_1$, $K \subset M_2$ are compact subsets then $O_{I \times K} = O_I \times O_K$. Suppose, further, that M_1', M_2' are the same as M_1, M_2 (respectively) and $f : M_1' \to M_1$, $h : M_2' \to M_2$ are continuous maps. Then $\deg_{I \times K}(f \times h) = (\deg_I f) \cdot (\deg_K h)$.*

1.2. Existence theorem

1.2.1.   The key to the extension results we are looking for is the following

Definition. *Let a topological group H act on a metric space E. Let $D_0 \subset E$ be open in its closure D. Then D is said to be a quasi-fundamental domain of the H-action on E if the following conditions are satisfied:*

(a) $H(D) = E$;

(b) $g(D_0) \cap h(D_0) = \emptyset$ $(g \neq h; g, h \in G)$;

(c) $E \backslash H(D_0) = H(D \backslash D_0)$.

If E is finite-dimensional and the following additional condition holds

(d) $\dim D = \dim E/H$; $\dim (D \backslash D_0) < \dim D$; $\dim H(D \backslash D_0) < \dim E$

   then D is said to be a fundamental domain.

Note, that if H is a discrete group then one can set D_0 to be the interior of D. Hence the definition above naturally complies with the classical one (cf. e.g. [DFN], p. 169).

1.2.2.

Theorem 1.1. *A (quasi)-fundamental domain exists for any free action of a compact Lie group G on a metric space X.*

First let us note the following technical fact.

Lemma 1.14. *For any locally finite open covering $\{U_\alpha\}_{\alpha \in I}$ of an n-dimensional metric space M there exists an open family $\{V_\alpha\}_{\alpha \in I}$ such that:*

1) $\overline{V}_\alpha \subset U_\alpha$ *for all* $\alpha \in I$;

2) $V_\alpha \cap V_\mu = \emptyset$ *if* $\alpha \neq \mu$;

3) $\dim \partial V_\alpha \leq n - 1$;

4) $\{\overline{V}_\alpha\}_{\alpha \in I}$ is a covering of M.

Proof of Lemma 1.14. According to Lemma 1.7 there exists an open covering $\{W_\alpha\}_{\alpha \in I}$ of M such that $\overline{W}_\alpha \subset U_\alpha$. Suppose (by induction) that $W_\alpha \subset Z_\alpha \subset \overline{Z}_\alpha \subset U_\alpha$ for every $\alpha \in I$, where Z_α is open and $\dim \partial Z_\alpha < \dim Z_\alpha$ (Lemma 1.3). Well-order the index set I and put $V_1 = Z_1$. Suppose that the sets V_α are defined for all $\alpha < \alpha_0$ and let

$$\Lambda = \bigcup_{\alpha < \alpha_0} V_\alpha \ .$$

It is easy to see that $\{V_\alpha\}_{\alpha \in I}$ is also locally finite, and hence (Lemma 1.8)

$$\partial \Lambda \subset \bigcup_{\alpha < \alpha_0} \partial Z_\alpha \ .$$

Let $\alpha_1 \in I$ be a first element such that Z_{α_1} is not contained in Λ. Set $V_{\alpha_0} = Z_{\alpha_1} \backslash \Lambda$. Since the sets $\overline{Z}_\alpha$ are canonically closed (see, [AP], p. 48), it follows from Lemma 1.8 that

$$\partial V_{\alpha_0} \subset \partial \Lambda \cup \partial Z_{\alpha_0} \subset \bigcup_{\alpha \leq \alpha_0} \partial Z_\alpha \ ,$$

and since the family $\{V_\alpha\}_{\alpha \in I}$ is locally finite it follows from Lemma 1.4 that $\dim \partial V_{\alpha_0} \leq n - 1$.

It is clear that conditions 1), 2) and 4) also hold and our lemma is proved by (maybe transfinite) induction.

1.2.3.   Proof of Theorem 1.1. According to Lemma 1.1, an orbit map $p : X \to X/G$ defines a fiber bundle $\rho = (X, p, X/G)$ with a fiber G. Let us consider an open covering $\{U_\alpha\}_{\alpha \in I}$ of X/G such that ρ is trivial over every U_α. Since X is metric and p is a closed map ([Bre], p. 38), the space X/G is paracompact ([En], Theorem 5.1.33). By Lemma 1.1 X/G is locally metrizable and hence metrizable as a locally metrizable paracompact space ([AP], p. 280). Thus one can assume that the covering $\{U_\alpha\}$ is locally finite. Let $\{V_\alpha\}_{\alpha \in I}$ be a refinement of $\{U_\alpha\}_{\alpha \in I}$ satisfying the conditions of Lemma 1.14 (of course, property 3) of the conclusion of this lemma is well-defined only under the assumption that X is finite-dimensional).

Suppose, further, that $h_\alpha : G \times U_\alpha \to p^{-1}(U_\alpha)$ is a homeomorphism, defining a chart in ρ over U_α, and set $h_\alpha^1 = h_\alpha | G \times V_\alpha$. Since the family $\{V_\alpha\}_{\alpha \in I}$ is disjoint (condition 2)) the homeomorphisms $\{h_\alpha^1\}_{\alpha \in I}$ induce (Lemma 1.2) a homeomorphism

$$h^1 : G \times \bigcup_\alpha V_\alpha \approx \bigcup_\alpha (G \times V_\alpha) \approx \bigcup_\alpha p^{-1}(V_\alpha) \ .$$

Set

$$D_0 = h^1 \left(e \times \bigcup_\alpha V_\alpha \right) \approx \bigcup_\alpha h^1(e \times V_\alpha), \ D = \overline{D}_0$$

and show that D satisfies the definition of the (quasi)-fundamental domain (here e is the identity of G).

Since p is the closed map

$$p(D) = p(\overline{D_0}) = \overline{p(D_0)} = X/G \; ,$$

condition (a) of the definition of the (quasi)-fundamental domain follows immediately. In order to prove (b) it suffices to note that for any g, $s \in G$ the following equalities hold:

$$g(D_0) = \bigcup_\alpha h^1(g \times V_\alpha); \quad s(D_0) = \bigcup_\alpha h^1(s \times V_\alpha) \; .$$

To verify condition (c) note that since $G(D\backslash D_0) \cup G(D_0) = X$ it suffices to prove that $G(D\backslash D_0) \cap G(D_0) = \emptyset$. Let $x \in G(D\backslash D_0) \cap G(D_0)$. Then there exist $z \in D_0$, $y \in D\backslash D_0$ and $g, h \in G$ such that $x = gy = hz$, i.e. y and z belong to the same orbit. But on the one hand,

$$y \in D\backslash D_0 \subset \bigcup_\alpha p^{-1}(\partial V_\alpha) = p^{-1}\left(\bigcup_\alpha \partial V_\alpha\right) \; ,$$

from which it follows that

$$p(y) \in p\left(p^{-1}\left(\bigcup_\alpha \partial V_\alpha\right)\right) = \bigcup_\alpha \partial V_\alpha \; .$$

On the other hand,

$$p(z) \in p(D_0) = X/G\backslash \bigcup_\alpha \partial V_\alpha \; ,$$

a contradiction.

It remains to show (d). Denote the set $h^1(e \times V_\alpha)$ by W_α. By the local finiteness of the family $\{U_\alpha\}_{\alpha \in I}$ and condition 1) the family $\{W_\alpha\}_{\alpha \in I}$ is also locally finite, hence (Lemma 1.4) $\dim D = \max_\alpha \{\dim \overline{V}_\alpha\}$. According to 1) $\overline{W}_\alpha$ is homeomorphic to $\overline{V}_\alpha$ from which it follows that $\dim D = \dim X/G$.

Using condition 3) and Lemma 1.5 we have:

$$\dim G(D\backslash D_0) = \dim\left(p^{-1}\left(\bigcup_\alpha \partial V_\alpha\right)\right)$$
$$\leq \dim G + \dim\left(\bigcup_\alpha \partial V_\alpha\right)$$
$$\leq \dim G + \dim(X/G) - 1$$
$$= \dim G + \dim X - \dim G - 1 = \dim X - 1 \; .$$

In order to complete the proof of (d) we note that by Lemma 1.8

$$D \backslash D_0 \subset \bigcup_\alpha (\overline{W}_\alpha \backslash W_\alpha)$$

and by Lemma 1.4

$$\dim \left(\bigcup_\alpha (\overline{W}_\alpha \backslash W_\alpha) \right) \leq \dim(X/G) - 1 \, ,$$

i.e. $\dim(D \backslash D_0) \leq \dim(X/G) - 1$.

Theorem 1.1 is proved.

1.3. Equivariant Kuratowski-Dugundji Theorem

1.3.1. In this section we shall study the following problem. Let X and Y be metric G-spaces and let $A \subset X$ be a closed invariant subspace. Assume also that G is a compact Lie group acting on $X \backslash A$ with a finite number of orbit types $(H_1), (H_2), \ldots, (H_m)$ (in short, X has a finite orbit type). What are the conditions on X and Y which guarantee that an equivariant map $f : A \to Y$ has an equivariant extension over X (over some invariant neighbourhood of A in X)?

First, let us recall the principle of constructing equivariant maps via induction over orbit types (see, for instance, [Di1], p. 203). Suppose that Or is a finite set of conjugacy classes of subgroups of G. We can choose an admissible indexing Or $= \{(H_1), (H_2), \cdots, (H_k)\}$, this meaning that $(H_j) < (H_i)$ implies $i < j$. If the G-space X has finite orbit type we always choose an admissible indexing of its set of orbit types $\mathrm{Or}(X)$. Let $f : X \to Y$ be an equivariant map between G-spaces of finite orbit type. Let

$$\mathrm{Or}(X) \bigcup \mathrm{Or}(Y) = \{(H_1), \cdots, (H_m)\}$$

be an admissible ordering. Define a filtration of X by closed invariant G-subspaces

$$X_1 \subset X_2 \subset \cdots \subset X_m = X, \tag{1.1}$$

where

$$X_i = \{x \in X \mid (G_x) = (H_j) \quad \text{for some} \quad j \leq i\}.$$

In a similar way one can define a filtration of Y:

$$Y_1 \subset Y_2 \subset \cdots \subset Y_m = Y. \tag{1.2}$$

We are interested in the existence of the equivariant extensions $\gamma : X_i \to Y_i$ of the equivariant maps $\delta : X_{i-1} \to Y_{i-1}$.

Lemma 1.15 (see [Di1], proposition 8.1.5). *Suppose that* $s : K^{H_i} \to Y^{H_i}$ *is a* $W(H_i)$-*equivariant extension of the map* $\delta|K_{i-1}^{H_i}$. *Then there exists a unique G-equivariant extension* $\gamma : K_i \to Y_i$ *such that* $\gamma|K^{H_i} = s$.

With the aid of Lemma 1.15 this problem is reduced to the following one: let X, Y be G-spaces and let $A \subset X$, $B \subset Y$ be closed G-subspaces such that the actions of G on $X\backslash A$ and $Y\backslash B$ are supposed to be free. It is necessary to extend the G-equivariant map $f : A \to B$ to a G-equivariant map $F : X \to Y$. Since $X\backslash A$ is a free G-space, the fiber bundle over $(X\backslash A)/G$ is given. Hence, one has to solve two problems: one is to extend f over some invariant neighbourhood U of A in X, and another is to extend a partial section of an associated bundle (cf. [Di1]). In this way a general problem of equivariant extension is usually reduced to a sequence of extensions of partial sections. This approach was utilized by J. Jaworowski [Ja1], [Ja2], R. Lashof [La], M. Madirimov [Mad1], [Mad2] (see also general discussion in [Di1], p. 205) and others.

We describe below another general approach which is intuitively easier and allows to obtain stronger extension results in certain cases. We hope that the applications of our approach will not be confined to the problems which are touched upon in this book.

1.3.2.   Let us assume that a compact Lie group G acts on a metric space X and let $A \subset X$ be a closed invariant subset such that the action of G on $X\backslash A$ is free. Set $X^{(0)} = X$, $L^{(0)} = X^{(0)}\backslash A$. By Theorem 1.1 there exists a fundamental domain $D^{(0)} \subset L^{(0)}$. Let $D_0^{(0)}$ be the corresponding open subset of $D^{(0)}$, satisfying the conditions (b) – (d) of the definition of the (quasi)-fundamental domain, and let $X^{(1)} = A \cup G(D^{(0)}\backslash D_0^{(0)})$, $L^{(1)} = X^{(1)}\backslash A$. Now, applying Theorem 1.1 to $X^{(1)}\backslash A$ we get $X^{(2)}$ and $L^{(2)}$, etc. So we have a closed invariant filtration

$$X = X^{(0)} \supset X^{(1)} \supset X^{(2)} \supset \cdots \supset A .$$

If $X\backslash A$ is finite-dimensional then this filtration is finite (condition (d)); in this case put $A = X^{(l)}$. Let Y be another G-space.

Proposition 1.1. *If $X\backslash A$ is finite-dimensional then any equivariant map $A \to Y$ extends over X if for all $i \geq 1$ any equivariant map $X^{(i)} \to Y$ has a (non-equivariant) extension over $X^{(i)} \cup D^{(i-1)}$. The same is true for extensions of equivariant homotopies.*

Proof of Proposition 1.1. We need the following

Lemma 1.16 ([Bre], p. 39). *Let Γ be a compact group acting on the spaces X and Y. Let $C \subset X$ be any closed subset and let $h : C \rightarrow Y$ be a map such that whenever c and gc are both in C (for some $g \in \Gamma$), then $hgc = ghc$. Then h can be extended uniquely to an equivariant map of $G(C)$ into Y.*

In order to prove the proposition we shall construct the desired map by induction. Suppose that $f_i : X^{(i)} \rightarrow Y$ is an equivariant extension of the given map. By the assumption there exists a continuous extension $\tilde{f}_i : X^{(i)} \cup D^{(i-1)} \rightarrow Y$ and the result follows from the properties of a fundamental domain and Lemma 1.16.

1.3.3.   Now with the help of Proposition 1.1 we shall prove an equivariant analogue of the Kuratowski-Dugundji Theorem. We keep in mind the following well-known

Theorem 1.2 ([Bor1], p. 80). *If a metric space T is locally k-connected for each $k = 0, 1, \ldots, n$, B is a closed subset of a metric space F and $\dim(F \backslash B) \leq n + 1$, then for every continuous map $f : B \rightarrow T$ there exists a neighbourhood V of the set B in F such that f has a continuous extension over V. If, in addition, T is globally k-connected for each $k = 0, 1, \ldots, n$ then f has a continuous extension over F.*

We shall prove

Theorem 1.3. *Suppose that X and Y are metric G-spaces, $A \subset X$ is a closed invariant subset, G is a compact Lie group acting on $X \backslash A$ with a finite number of orbit types $(H_1), (H_2), \ldots, (H_m)$. Suppose, further, that the set Y^{H_i} is locally and globally k-connected for each $k = 0, 1, \ldots, n_i$ $(i = 1, \ldots, m)$. If $\dim(X_{(H_i)} \backslash A)/G \leq n_i + 1$ for every i then any equivariant map $f : A \rightarrow Y$ has an equivariant extension over X. If for each $k = 0, 1, \ldots, n_i$ the set Y^{H_i} is locally k-connected $(i = 1, \ldots, m)$, and a cone over Y is metrizable, then f has an equivariant extension over some invariant neighbourhood of A in X.*

(Note, that a cone over Y is metrizable if Y is compact).

Proof of Theorem 1.3. Let us show, to begin with, how to deduce the local variant of Theorem 1.3 from the global one. Consider a cone $K(Y)$ over Y with the conical action of G (see, for instance, [Bre], p. 99) defined as follows:

$$g(x, t) = \begin{cases} (gx, t) & \text{if } t \in (0, 1] \\ y_0 & \text{if } t = 0 \end{cases}$$

(here y_0 is the vertex of the cone). It is obvious that $(K(Y))^{H_i} = K(Y^{H_i})$, hence $(K(H))^{H_i}$ is locally and globally k-connected for each $k = 0, 1, \ldots, n_i$, where $i = 1, \ldots, m$. Assume that for an equivariant map $f : A \rightarrow Y \subset K(Y)$ there exists an equivariant extension $\tilde{f} : X \to Y$. Set $U = \tilde{f}^{-1}(K(Y)\backslash\{y_0\})$ and note that the composition of $f|U$ on U with the natural retraction of the set $K(Y)\backslash\{y_0\}$ onto Y (along the rays) is a local extension of f.

The global version of the theorem follows by induction over the orbit types from Proposition 1.1 and Lemma 1.15. For completeness we present some details in the following

Lemma 1.17. *Suppose that $X\backslash A$ is a free G-space, Y is a locally and globally k-connected G-space for each $k = 0, 1, \ldots, n$ and $f : A \to Y$ is an equivariant map. If $\dim(X\backslash A)/G \leq n+1$ then f has an equivariant extension over X.*

Proof of Lemma 1.17. Let $D^{(j)}, D_0^{(j)}, X^{(j)}$ and $L^{(j)}$ be as in Proposition 1.1. According to condition (d) of the definition of the fundamental domain we have: $\dim D^{(j-1)} = \dim\left(L^{(j-1)}/G\right) \leq \dim\left(L^{(0)}/G\right) = \dim((X\backslash A)/G) \leq n + 1$ for every $j < l$. Hence by Theorem 1.2 there exists a (non-equivariant) extension $\tilde{f}_j : X^{(j)} \cup D^{(j-1)} \to Y$ of f. In order to finish the proof, it remains to use Proposition 1.1. So the proof of the lemma (and hence of Theorem 1.3) is completed.

1.3.4.   Here we consider some corollaries of Theorem 1.3.

Corollary 1.1. *Suppose that under the conditions of the global (local) variant of Theorem 1.3 $\dim\left(X_{(H_i)}\backslash A\right)/G \leq n_i$ for every $i = 1, \ldots, m$. Then every equivariant homotopy $\varphi : A \times [0,1] \to Y$ has an equivariant extension over X (over some invariant neighbourhood of A in X).*

The proof is obvious.

Let $f : B \to T$ be a continuous map and N be a subspace of B. A homotopy $h : N \times [0,1] \to T$ is said to be a partial homotopy of the map f, if $f|N = h|N \times \{0\}$ (see [Hu], p. 13). A homotopy $\mu : B \times [0,1] \to T$ is said to be an expansion of the homotopy $h : N \times [0,1] \to T$, if $\mu|N \times [0,1] = h$.

Corollary 1.2 (cf. [Hu], p. 14). *Under the conditions of the local variant of Corollary 1.1 every partial equivariant homotopy $\varphi : A \times [0,1] \to Y$ of the equivariant map $f : X \to Y$ has an equivariant expansion over X.*

Proof of Corollary 1.2. Let us define an equivariant map $h : A \times [0,1] \cup X \times \{0\} \to Y$ by the formula:

$$h(x,t) = \begin{cases} f(x) & \text{if } x \in X, \ t = 0 \\ \varphi(x,t) & \text{if } x \in A, \ t \in [0,1] \ . \end{cases}$$

By the assumption h has an equivariant extension $\widetilde{h} : V \to Y$, where V is an invariant neighbourhood of the set $A \times [0,1] \cup X \times \{0\}$ in $X \times [0,1]$. Since the segment $[0,1]$ is compact there exists an invariant neighbourhood W of A in X such that $W \times [0,1] \subset V$; set $\overline{h} = \widetilde{h}|W \times [0,1] \cup X \times \{0\}$. By Urysohn's Lemma there exists a continuous function $\alpha : X/G \to [0,1]$ such that $\alpha((X \backslash W)/G) = 0$ and $\alpha(A/G) = 1$. We shall define (cf. [Hu], p. 14) the required map $F : X \times [0,1] \to Y$ by setting $F(x,t) = \overline{h}(x, \alpha(\overline{x}) \cdot t)$, where $\overline{x}$ is an image of x under the orbit map $p : X \to X/G$. The equivariance of the constructed map is obvious.

Using the Mostov Embedding Theorem [Mo] one can easily prove (with the help of Theorem 1.3) the following result of J. Jaworowski [Ja1]).

Corollary 1.3. *Suppose that a compact Lie group G acts on a locally compact, separable, finite-dimensional metric space X with a finite number of orbit types. Then X is a $G - ENR$-space iff X^H is an ENR-space for every $H \in \mathrm{Iso}(X)$.*

From the global variant of Theorem 1.3 it follows immediately

Corollary 1.4.   *Let a finite group G act on a compact manifold X^n, let $A \subset X^n$ be a closed invariant subset and let the action of G on $X^n \setminus A$ be free. Let V be an orthogonal $(n + k)$-dimensional representation of G ($k \geq 1$) and B a unit ball bounded by the unit sphere $S(V) = \partial B$. Then any equivariant map $f : A \to S(V)$ has an equivariant extension $F : X^n \to S(V)$.*

Corollary 1.5.   *Let a finite group G act on a compact smooth manifold X^n, let $K \subset X^n$ be a closed invariant subset and let the action of G on $X^n \setminus K$ be free. Let V be an orthogonal n-dimensional representation of G and $B \subset V$ a unit ball in V. Then any equivariant map $f : K \to B \setminus \{0\}$ has an equivariant extension $F : X^n \to B$ such that the set $F^{-1}(0)$ is finite.*

To prove this result we need the following

Lemma 1.18 (see [Di1, Bre]). *Let a finite group G act on a compact smooth manifold X^n, let $K \subset X^n$ be a closed invariant subset and let the action of G on $X^n \setminus K$ be free. Then any neighborhood of K contains an open invariant neighborhood U such that $X \setminus U$ is a compact smooth manifold with boundary. Moreover, $X \setminus U$ possesses a finite invariant triangulation.*

Proof of Corollary 1.5. Take a local equivariant extension $\overline{f} : \overline{U} \to B \backslash \{0\}$ of f provided by Theorem 1.3 (here $\overline{U}$ is an invariant neighbourhood of K). Let $U \subset \overline{U}$

be an invariant neighborhood of K provided by Lemma 1.18. Using Theorem 1.3 once again extend $\bar{f}|U$ to an equivariant map $\mu : U \cup X^{n-1} \to B \setminus \{0\}$ (here X^{n-1} is an invariant $(n-1)$-skeleton in X). To complete the proof take a "simplicial" fundamental domain $D \subset (X \setminus (X^{n-1} \cup U))$, extend μ (non-equivariantly) to a map $\nu : U \cup X^{n-1} \cup D \to B$ with a finite number of "zeros" and use Lemma 1.16.

1.3.5.   Below we will introduce certain notions related to filtrations (1.1), (1.2) and Theorem 1.3. These notions will play an important role in Chapter 3.

Let X, Y and G be as in Subsection 1.3. Denote filtrations (1.1) and (1.2) by F_X and F_Y respectively.

Definition.   *We say that the pair (X, Y) is (G, F_X, F_Y)-extendable if the following conditions hold:*

a) there exists an equivariant map $f : X \to Y$;

b) for any equivariant map $X_i \to Y_i$ there exists an equivariant extension over X.

It is easy to construct an example of the situation when an equivariant map $X \to Y$ exists but the pair (X, Y) is not (G, F_X, F_Y)-extendable.

Indeed, let V and W be finite-dimensional spaces with $\dim V = \dim W$. Denote by $S(V)$ and $S(W)$ unit spheres in V and W respectively. Let, further, $V = V_1 \oplus V_2 \oplus V_3$ and $W = W_1 \oplus W_2 \oplus W_3$. Let p be a prime and $G = Z_{p^2}$ act with three orbit types on $S(V)$ and $S(W)$: trivially on V_1 and W_1, freely on V_3 and W_3 and as Z_p on V_2 and W_2. Suppose that $\dim V_1 = \dim W_1$ and $\dim V_2 > \dim W_2$.

Of course, there exists an equivariant map $j : S(V) \to S(W)$, namely, $j(x) \equiv y \in S(W^G)$. However, any map $i : S(V_1) \to S(W_1)$ such that $\deg i \not\equiv 0 \pmod{p}$ (e. g. induced by an isomorphism of the vector spaces) cannot be extended equivariantly onto $S(V_2)$. Indeed, let f be such an extension. There exists an equivariant map $g : S(W_2) \to S(V_2)$, such that $\deg g|S(W_1) \not\equiv 0 \pmod{p}$ (by Theorem 1.3 any map from $S(W_1) \to S(V_1)$ extends equivariantly over W_2). Take the map $g \circ f = h : S(V_2) \to S(V_2)$. By construction and Corollary 3.2 (cf. Chapter 3) we must have $0 = \deg h \equiv \deg h|V_1 \pmod{p}$. However, $h|V_1 = (g|W_1) \circ i$, therefore $\deg h|V_1 \not\equiv 0 \pmod{p}$.

It follows immediately from Theorem 1.3

Corollary 1.6.   *Let (X, Y) be a pair of G-spaces with G-invariant filtrations F_X and F_Y as above. If there exists an equivariant map $X_1 \to Y_1$ and for any $x \in X \setminus X_1$ the set Y^{G_x} is $(\dim(X^{G_x}) - \dim W(G_x))$-connected then (X, Y) is (G, F_X, F_Y)-extendable.*

Remark 1.3.   We don't know if Corollary 1.6 describes all the (G, F_X, F_Y)-extendable pairs. Observe also that if the pair is not (G, F_X, F_Y)-extendable then the "current" step of the induction depends on the previous one, and it might happen that after several extensions we will be "stuck" although an equivariant map $X \to Y$ does exist. Roughly speaking, this is the case when the first obstruction is not enough (cf. Corollary 1.6). When one studies the equivariant homotopy classification problem (EHCP) for equivariant maps of spheres of different dimensions, the necessity to struggle with the above effect appears in the natural way (see, for instance, [IV, IMV1, IMV2] where some results for abelian group linear representations are given). However, in this book we study either the EHCP in the case of the "first obstruction" (see Chapter 5) or the Brouwer degrees of equivariant maps of manifolds having the same dimensions (see Chapters 2 and 3). In the second chapter we consider topological (non-smooth) actions in the "first obstruction type" situation. In Chapter 3 assuming the actions to be smooth we will show how certain "ad hoc" arguments allow to go around the problem of the "second obstruction" and to get the degree results without any "obstruction/dimension restrictions". We are grateful to J. Ize who pointed out the importance of studying the "extendability problem" in the context relevant to our discussion.

Consider now a more general situation:

Let X, Y, G, F_X and F_Y be as above, let K be a subgroup of G and let $(\bar{X}, \bar{Y})$ be a K-invarinat pair ($\bar{X} \subset X, \bar{Y} \subset Y$).

Definition.   We will say that $(\bar{X}, \bar{Y})$ is (K, F_X, F_Y)-extendable pair if conditions a) and b) from the above definition are satisfied by filtrations $\{\bar{X}_i = X_i \cap \bar{X}\}$, $\{\bar{Y}_i = Y_i \cap \bar{Y}\}$ and K-equivariant maps.

In Chapter 3 we will use the above definitions to distinguish those orbit types which actually participate in the comparison principle formula.

1.4. Historical and bibliographical notes

1.4.1. All the undefined notions and standard facts from G-space theory and fiber bundle theory can be found in [Bre], [Di2] and [Hus], from general topology and dimension theory – in [En], from retract theory – in [Bor1], from algebraic topology – in [Do1], from obstruction theory – in [Hu] and [Di2].

1.4.2. In the context relevant to our discussion, a construction of a simplicial fundamental domain was used in [Ei]. For a free action of a cyclic group on a sphere Theorem 1.1 was obtained by M. Krasnoselskiı [Kr1] (see also [KZ]); an analogous construction for a free action of a cyclic group on an invariant domain of a sphere was offered by P. Zabrejko [Za1,Za2]. A case of a finite group action on a subspace of R^n was considered by Z. Balanov and S. Brodsky [BB1,BB2,Ba1]. In full generality, Theorem 1.1 was proved by Z. Balanov and A. Kushkuley [BK1,BK2,KB,Ba2,BKZ1, BKZ2]. For abelian group linear actions "fundamental cells" were studied in the recent paper by J. Ize and A. Vignoli.

1.4.3. Under the assumptions that all Y^{H_i} are ANR-spaces (correspondingly, AR-spaces) and X is locally compact and separable, Theorem 1.3 was proved by J. Jaworowski [Ja1]. R. Lashof [La] eliminated a condition of local compactness of X; M. Madirimov [Mad1,Mad2] removed a condition of separability of X and weakened the requirements on Y^{H_i}. In the given formulation Theorem 1.3 was proved by Z. Balanov and A. Kushkuley [BK1,BK2,KB,Ba2,BKZ1,BKZ2]. In contrast to [Ja1,La,Mad1,Mad2] we did not use the embedding theorem of Mostov (see also [Ja2]). The existence of equivariant extensions for abelian group linear actions was studied in the recent paper [IV].

Our Corollary 1.1 is an equivariant analogue of Borsuk's Theorem about homotopy extension [Bor1], p. 80; Corollary 1.2 is an equivariant analogue of Borsuk's Theorem about homotopy expansion [Bor1], p. 94.

Chapter 2

DEGREE THEORY FOR EQUIVARIANT MAPS

OF FINITE-DIMENSIONAL MANIFOLDS:

TOPOLOGICAL ACTIONS

The equivariant Kuratowski-Dugundji Theorem proved in the first chapter is the main tool in considerations of this chapter. In the first section we prove the general comparison formula for degrees of maps which are equivariant with respect to finite group actions. In the second section we are interested in some special cases: free actions of compact Lie groups, (non-free) actions of p-groups and some others. In the third section we give some counterexamples.

2.1. Comparison principle for a finite group

2.1.1. It is well known that the degree of an equivariant map from a manifold M to a sphere S ($\dim M = \dim S$) is not completely arbitrary as it is in the non-equivariant case covered by the Hopf Theorem. For example, if a finite group G acts freely on M then for any G-equivariant map $f : M \to S$, $\deg f$ is unique modulo $|G|$. Thus we naturally arrive at the necessity to consider the following problem:

Let a compact Lie group G act on a topological (compact, connected, oriented) n-dimensional manifold M and on the oriented n-dimensional sphere S. Suppose that the actions of G on M and S are concordant (see Remark 1.1(a)). What are the restrictions on possible values of the degrees of equivariant maps?

The following two approaches are the most essential in the investigation of this problem among the numerous ones. In the first of them (the homological approach) the degree of an equivariant map is expressed in terms of the homological characteristics of the corresponding actions (e.g. the so-called Smith indices) and in terms of the homological characteristics of the restrictions of maps in question to the relevant sets of fixed points. P. Smith [Le], p. 350, was the initiator of this approach. In the second approach (the so-called geometric approach) the obstruction to the existence of an equivariant homotopy between a couple of the equivariant maps is evaluated with the help of the relevant equivariant extension theorems. The idea behind this approach can be traced back to the original papers of S. Eilenberg [Ei] and M. Krasnoselskii [Kr1].

The aim of this section is to prove (in the framework of the geometric approach) the following

Theorem 2.1. *Let G be a finite group acting on M and S, let $N \subset M$ be a closed invariant subset, let $(H_1), (H_2), \ldots, (H_m)$ be orbit types in $M \backslash N$. Assume also that the set S^{H_i} is locally and globally k-connected for all $k = 0, 1, \ldots, \dim M^{H_i} - 1$ where $i = 1, \ldots, m$. Then for every pair of equivariant maps $\Phi, \Psi \colon M \to S$ which are equivariantly homotopic on N the following formula is true:*

$$\deg \Psi \equiv \deg \Phi \quad (\mathrm{mod}\, GCD\{|G/H_1|, \ldots, |G/H_m|\}) \ .$$

It is more convenient for us to prove this theorem in the following equivalent form.

Theorem 2.1'. *Under the assumptions of Theorem 2.1 there exist integers $a_1, \ldots, a_m$ such that*

$$\deg \Psi - \deg \Phi = \sum_{i=1}^{m} a_i \cdot |G/H_i| \ .$$

Proof of Theorem 2.1'. Let us define a cylindric action of a group G on the cylinder $C = M \times [0, 1]$ by setting it to be trivial on the segment $[0, 1]$. We can also define a conic action on the ball B bounded by sphere S. Let O be the center of the ball. For equivariant maps $\Phi, \Psi \colon M \to S$ an equivariant map $f_0 \colon M \times \{0, 1\} \bigcup N \times [0, 1] \to B \backslash \{O\}$ is obviously defined. Let $f \colon C \to B$ be an equivariant extension of f_0 (see Theorem 1.3). Set $K = f^{-1}(O)$. If orientations on C and B are properly chosen then there exist fundamental classes $O_K \in H_n(C, C \backslash K)$ and $O_o \in H_n(B, B \backslash \{O\})$ which determine the degree of f by the formula: $f_*(O_K) = (\deg f) \cdot O_o$ (see Subsection 1.1.4)

Lemma 2.1. *Under the assumptions above $\deg f = \pm(\deg \Psi - \deg \Phi)$.*

This lemma follows immediately from Lemma 1.9.

If $m = 1$, i.e. there is only one orbit type $(H_1) = (H)$ outside N, we can proceed in the following way. As in Proposition 1.1, there exists a closed $W(H)$-invariant filtration

$$C^H = C^{(0)} \supset C^{(1)} \supset C^{(2)} \supset \cdots \supset (M \times \{0, 1\} \bigcup N \times [0, 1])^H \ .$$

By our assumptions $B^H\backslash\{O\}$ is , say, locally and globally k-connected for each $k = 0, 1, \cdots, s$, and the length of our filtration is less than s. Therefore, by Theorem 1.1 there exists a $W(H)$-equivariant extension $F_0 : C^{(1)} \to B^H\backslash\{O\}$ of the map f_0. It is clear now that F_0 can be further extended to the map $F : C^H \to B^H$ in such a way that

$$F^{-1}(O) = \{\text{disjoint union of } |W(H)| \text{ compact sets which are permuted by } W(H)\}.$$

Finally, using Lemma 1.15 and properties of the degree stated in Lemmas 1.10 – 1.12 we see that $\deg F$ is divisible by $|G/H|$. This is essentially the result obtained in [Ba1, Ba2, BB1, BB2] (and for smooth actions in [Bow1, Bow2]). In the general case ($m > 1$) one can try to use the standard induction over the orbit types described in Lemma 1.1 in order to obtain a map $F : C \to B$ with separated set of " zeros ". However, an extension constructed at the first step (the one we have just described) is no longer a map into $B\backslash\{O\}$ and the arguments presented above can not be used to make an induction step. Therefore, before proceeding with the induction step we will " build a wall" which separates zeros obtained on previous steps. This procedure is formalized below.

Lemma 2.2. *There exists an equivariant extension $F : C \to B$ of the map f_0 satisfying the following conditions:*

(α) $K = F^{-1}(O) = \bigcup\limits_{j=1}^{m} T_j$, $T_s \cap T_p = \emptyset$ *if* $s \neq p$;

(β) $T_j = G(K_j)$ *for some compact* K_j;

(γ) $H_j(K_j) = K_j$;

(δ) $g(K_j) \cap h(K_j) = \emptyset$ *if* $gh^{-1} \notin H_j$ $(j = 1, \ldots, m)$.

2.1.2.   Proof of Lemma 2.2. Define a "cylindric" filtration $\{C_i = M_i \times [0,1]\}$ of the G-space C and a "conic" filtration $\{B_i = K(S_i)\}$ of the G-space $K(S)$ as it was constructed in subsection 1.1 (see filtrations (1.1) and (1.2)). We will construct the map F using induction over the orbit types. We have already explained what happens at the base of the induction. In order to make the induction step one has to solve the following problem:

Let $\widetilde{f}_i : C_i \to B_i$ be an equivariant map $(i = 0, 1, \ldots, m - 1)$ such that $\widetilde{f}_i|(C_i \cap \partial C) = f_0$ and the set of "zeros" of $\widetilde{f}_i$ satisfies the following conditions:

(α') $\widetilde{f}_i^{-1}(O) = \bigcup\limits_{j=1}^{i} T_j^i$, where $T_s^i \cap T_p^i = \emptyset$ if $s \neq p$;

(β') $T_j^i = G(K_j^i)$ for some compact K_j^i;

(γ') $H_j(K_j^i) = K_j^i$;

(δ') $g(K_j^i) \cap h(K_j^i) = \emptyset$ if $gh^{-1} \notin H_j$.

(Here the upper index means a number of the induction step and the lower one – the number of the orbit type in $\mathrm{Or}(M_i \backslash N)$). Define a map

$$f_i : C_i \cup (\partial C \cap C_{i+1}) \to B_{i+1}$$

by the formula

$$f_i(x) = \begin{cases} \widetilde{f_i}(x) & \text{if } x \in C_i \ ; \\ f_0(x) & \text{if } x \in \partial C \cap C_{i+1} \ . \end{cases}$$

It is required to construct a G-equivariant extension $\widetilde{f}_{i+1} : C_{i+1} \to B_{i+1}$ with the properties (α') - (δ') (with $i + 1$ instead of i).

Let us consider the set $C_{i+1}^{H_{i+1}} = C^{H_{i+1}} \subset C_{i+1}$ and define $V_{i+1} = C_i^{H_{i+1}} \cup (\partial C \cap C^{H_{i+1}})$. It is clear that the restriction $f_i|V_{i+1}$ is $W(H_{i+1})$-equivariant. We want to construct a $W(H_{i+1})$-equivariant extension $f^{H_{i+1}} : C^{H_{i+1}} \to B^{H_{i+1}}$ of the map $f_i|V_{i+1}$ with the "nice" set of zeros.

By induction there exists a G-invariant small neighbourhood $U(T_j^i)$ of the set T_j^i ($j \le i$) such that:

1) $U(T_j^i) = \bigcup\limits_{s=1}^{|G/H_j|} g_s U(K_j^i)$, where $\{g_s\}_{s=1}^{|G/H_j|}$ are coset representatives of G/H_j;

2) $U(K_j^i)$ are H_j-invariant neighbourhoods of the sets K_j^i respectively;

3) $g_l U(K_j^i) \cap g_r U(K_j^k) = \emptyset$ if $l \ne r$;

4) $U(K_t^i) \cap U(K_p^i) = \emptyset$ if $t \ne p$.

Set $\widetilde{U}(T_j^i) = U(T_j^i) \cap C^{H_{i+1}}$; it is clear that $\widetilde{U}(T_j^i)$ is $W(H_{i+1})$-invariant. Let $V(T_j^i)$ be a closed $W(H_{i+1})$-invariant neighbourhood of T_j^i contained in $\widetilde{U}(T_j^i)$ such that $\dim \partial V(T_j^i) < \dim C^{H_{i+1}}$ (Lemma 1.6). Set

$$U^{(i)} = \bigcup\limits_{j=1}^{i} V(T_j^i) \ .$$

Then

$$W(H_{i+1})(U^{(i)}) = U^{(i)} \ ; \tag{2.1}$$

$$\dim(\partial U^{(i)}) < \dim C^{H_{i+1}} \ . \tag{2.2}$$

Taking into account (2.1) and the fact that $W(H_{i+1})$ acts freely on $C^{H_{i+1}} \backslash C_i^{H_{i+1}}$, we conclude that $W(H_{i+1})$ acts freely on $\left(\partial U^{(i)} \backslash \left(\partial U^{(i)} \cap V_{i+1}\right)\right) \subset C^{H_{i+1}}$. By our assumption the sets $\{B^{H_{i+1}} \backslash \{O\}\}$ are locally and globally k-connected for each $k = 0, 1, \ldots, \dim \partial U^{(i)} - 1$ (see (2.2)). Hence by Theorem 1.3 there exists a $W(H_{i+1})$-equivariant extension $\widehat{f}_i : V_{i+1} \cup \partial U^{(i)} \to B^{H_{i+1}}$ of the map $f_i|V_{i+1}$ without new zeros. It is no less clear that (e.g. by Theorem 1.3) there exists a $W(H_{i+1})$-equivariant extension $s_i : V_{i+1} \cup U^{(i)} \rightarrow B^{H_{i+1}}$ of the map $\widehat{f}_i$. Finally, the set $C^{H_{i+1}} \backslash \left(V_{i+1} \cup U^{(i)} \right)$ is a free $W(H_{i+1})$-subspace of the $W(H_{i+1})$-space $C^{H_{i+1}}$. As was explained above , there exists a filtration by closed $W(H_{i+1})$-invariant subsets

$$C_{i+1}^{(0)} = V_{i+1} \cup U^{(i)} \subset C_{i+1}^{(1)} \subset C_{i+1}^{(2)} \subset \cdots \subset C_{i+1}^{(l)} = C^{H_{i+1}}$$

of the space $C^{H_{i+1}}$ such that

$$\dim \left(C_{i+1}^{(r)} \backslash C_{i+1}^{(0)} \right) < \dim \left(C_{i+1}^{(r+1)} \backslash C_{i+1}^{(0)} \right)$$
$$\leq \dim \left((M \backslash N)^{H_{i+1}} \right) + 1$$
$$= \dim \left(M_{(H_{i+1})} \right) + 1 \ (r = 0, 1, \ldots, l - 1) .$$

Using Theorem 1.3, once again one can extend s_i to a $W(H_{i+1})$-equivariant map $f_i' : V_{i+1} \cup U^{(i)} \cup C_{i+1}^{(l-1)} \rightarrow B^{H_{i+1}}$ in such a way that all the zeros of f_i' belong to $U^{(i)}$. Further, there exists a (non-equivariant) continuous extension $f_i'' : V_{i+1} \cup U^{(i)} \cup C_{i+1}^{(l-1)} \cup D_{i+1}^{(l)} \rightarrow B^{H_{i+1}}$ of the map f_i' (here $D_{i+1}^{(l)}$ is a fundamental domain for the free $W(H_{i+1})$-space $C^{H_{i+1}} \backslash \left(V_{i+1} \cup U^{(i)} \cup C_{i+1}^{(l-1)} \right)$). Denote by L_{i+1} a compact set of zeros of the map $f_i'' | D_{i+1}^l$ and extend (by Lemma 1.16) the map f_i'' to a $W(H_{i+1})$-equivariant map $f^{H_{i+1}} : C^{H_{i+1}} \rightarrow B^{H_{i+1}}$. It is easy to see that the set of zeros in $C^{H_{i+1}} \backslash (V_{i+1} \cup U^{(i)})$ has the desired form $W(H_{i+1})(L_{i+1})$, where $h(L_{i+1}) \cap g(L_{i+1}) = \emptyset$ if $h \neq g$ $(h, g \in W(H_{i+1}))$.

Now, using Lemma 1.15, extend the $W(H_{i+1})$-equivariant map $f^{H_{i+1}}$ to a G-equivariant map $\widetilde{f}_{i+1} : C_{i+1} \rightarrow B_{i+1}$ and put

$$K_{i+1}^{i+1} = L_{i+1}, \ K_j^{i+1} = K_j^i \cup \left(\widetilde{f}_{i+1}^{-1}(O) \right) \cap U(K_j^i)), \ T_j^{i+1} = G\left(K_j^{i+1} \right) \ (j = 1, 2, \ldots, i) .$$

It is clear from the construction that the set $\widetilde{f}_{i+1}^{-1}(O)$ satisfies properties (α') – (δ') and the proof of Lemma 2.2 is complete.

2.1.3.   Continuation of the proof of Theorem 2.1. By Remark 1.1(a) we can assume without loss of generality that the actions of G on M and S are concordant.

Let $F : C \rightarrow B$ be a G-equivariant extension of the map f_0 satisfying the conclusion of Lemma 2.2. By Lemma 2.1

$$\deg F = \deg F | M \times \{0\} \cup M \times \{1\} = \pm(\deg \Psi - \deg \Phi) .$$

From Lemma 1.10 and condition (α) we have

$$\deg F = \sum_{j=1}^{m} \deg F_j ,$$

where F_j is a restriction of F to a sufficiently small neighbourhood of T_j. According to conditions (β) - (δ) every T_j may be represented as a union of disjoint sets K_{j_s} $(s = 1, 2, \ldots, |G/H_j|)$, so that every K_{j_s} is a translation of K_j. Hence, by of Lemma 1.10

$$\deg F_j = \sum_{s=1}^{|G/H_j|} \deg F_{j_s} ,$$

where F_{j_s} is the restriction of F_j to a sufficiently small neighbourhood of K_{j_s}. Now using Lemmas 1.11 and 1.12, G-equivariance of F, and taking into account that the actions of G on M and S are concordant (see Remark 1.1(a)), we can conclude that for every $j = 1, \ldots, m$ all $\deg F_{j_s}$ are equal to the same number a_j $(s = 1, \ldots, |G/H_j|)$. Thus $\deg F_j = a_j \cdot |G/H_j|$, from which the conclusion of Theorem 2.1' follows immediately.

2.2. Some special cases

2.2.1.   Here we shall consider some consequences from Theorem 2.1.

Corollary 2.1.   *Suppose that under the assumptions of Theorem 2.1 M and S coincide as the G-spaces, Ψ is the identity map and $N \supset S^G$. Then*

$$\deg \Phi \equiv 1 \pmod{GCD\{|G/H_i|\}} .$$

We continue to assume all the conditions of Theorem 2.1 throughout the rest of this subsection. We also assume that the action of G on M is effective.

Corollary 2.2.   *Let G be a p-group and let $N \supset M^G$ (e.g. G acts on M without G-fixed points), then*

$$\deg \Phi \equiv \deg \Psi \pmod{p} .$$

Corollary 2.3.   *Let $G = T$ be a torus and let $N \supset M^G$ (e.g. T acts on M without T-fixed points), then*

$$\deg \Phi = \deg \Psi.$$

Proof of Corollary 2.3. Since the set of finite p-subgroups is dense in T one can find a p-subgroup H in T such that $N \supset M^H = M^T$. Hence, by Corollary 2.2

$$\deg \Phi \equiv \deg \Psi \pmod{p} .$$

To complete the proof note that the prime p is arbitrary.

Corollary 2.4.   If the action of G on $M \setminus N$ is free then

$$\deg \Phi \equiv \deg \Psi \pmod{|G|}$$

if G is finite, and

$$\deg \Phi = \deg \Psi$$

if G is infinite.

Now let $N = \emptyset$ and let $S^{H_i} = S$ for all $i = 1, \ldots, m$. Define $M_{m-1} = \{x \in M : (G_x) = (H_j) \text{ for some } j \leq m-1\}$. It is well-known (see [Bre], p. 179) that $M \setminus M_{m-1}$ is open and everywhere dense in M, hence $\dim M_{m-1} < \dim M$. Then Theorem 1.3 implies that all equivariant maps from M to S are equivariantly homotopic on M_{m-1} and from Theorem 2.1 it follows

Corollary 2.5.   Under the assumption above,

$$\deg \Phi \equiv \deg \Psi \pmod{|G/H_m|} .$$

From this formula it follows immediately

Corollary 2.6.   Let the action of G on S be trivial. Then for any continuous map $\Phi : M \to S$ which is constant on the orbits one has

$$\deg \Phi \equiv 0 \pmod{|G/H_m|} .$$

2.2.2.   Here we give two consequences from Theorem 2.1 based on a trick due to
A. Dold [Do2] (see also Section 5.4).

In his classical work [Le], P. Smith established the following formula for the
degree of a map which is equivariant with respect to a pair of free actions of a cyclic
group of prime order p on homological spheres:

$$\deg f = (\mathrm{ind}(Z_p)_1)^{-1} \cdot (\mathrm{ind}(Z_p)_2) \,, \tag{2.3}$$

where $\mathrm{ind}(Z_p)_i$ $(i = 1, 2)$ are the so-called Smith indices of the first and second
actions and formula (2.3) is written in Z_p. Moreover, the Smith indices are non-
zero modulo p and hence formula (2.3) implies that the degree of an equivariant
map must be non-zero modulo p. In order to prove formula (2.3) P. Smith used
methods which are rather technical in nature. On the other hand, using Theorem
1.3 and Corollary 2.1 one easily obtains

Corollary 2.7. *Let a compact Lie group G act freely on a compact, oriented,
topological manifold M^m and the sphere S^k. Suppose that an equivariant map
$f : S \to M$ exists. Then $k \leq m$, and if $m = k$ then $\deg f$ is relatively prime to $|G|$
for finite G and equals ± 1 for infinite G.*

Proof of Corollary 2.7. If $m \leq k$ then by Theorem 1.3 there exists an equivariant map $h : M \to S$ and by Corollary 2.1 the degree of the map $h \circ f : S \to S$ equals
one modulo $|G|$ if G is finite. Further, note that any infinite compact Lie group
contains a torus and hence it contains finite subgroups of arbitrary large order.

The same argument also yields

Corollary 2.8.   *Let M, S and G be as in Theorem 2.1 and let $N = \emptyset$. Suppose
that the sets S^{H_i} are locally and globally t-connected for each $t = 0, 1, \cdots, \dim S^{H_i} -
1$. Suppose also that $s = GCD\{|G/H_i|\} > 1$. Then the degree of an equivariant
map $S \to M$ must be non-zero modulo s.*

Note that a condition $s > 1$ is obviously satisfied if G is a p-group which acts
without G-fixed points on S.

2.2.3.   In this subsection we present one corollary from Theorem 2.1 for infinite
group actions.

Let Γ be some group. A number $\pi(\Gamma)$ denotes the LCM of orders of all finite
subgroups of group G, if this LCM exists, and 0 otherwise. It is easy to construct an
infinite group G for which $\pi(G) > 0$; hence the following lemma is very interesting
in many cases.

Lemma 2.3. *Let the topological group G be compact and not completely discon-nected. Then $\pi(G) = 0$. (In particular, $\pi(G) = 0$ if G is an infinite compact Lie group).*

Proof of Lemma 2.3. By the Peter-Weyl Theorem ([Po], Theorem 33) the group G may be approximated by Lie groups, i.e. G is topologically isomorphic to a closed subgroup of a Cartesian product of Lie groups. If the image of G is finite under the projection of G on every component of this product, then the topology on G is completely disconnected. Otherwise, there is a continuous surjective homomorphism of G on some compact Lie group which contains a one-parameter subgroup (see, for instance, [Po], Theorem 58). Hence (see [Po], p. 332) G also contains a one-parameter subgroup. The closure of this subgroup is a connected compact abelian group, which is ([Po], Theorem 49) a torus and hence contains elements of arbitrary large orders. This completes the proof.

Return to the comparison principle. Denote by Fix (G, M) the union of all non-principal orbits for the action of G on M.

Corollary 2.9. *Suppose that equivariant maps Φ, $\Psi : M \to S$ are equivariantly homotopic on the set $\mathrm{Fix}(G, M)$. Then $\deg \Phi \equiv \deg \Psi \pmod{\pi(G)}$. In particular, if G is (algebraically) isomorphic to a compact and not completely disconnected group then $\deg \Phi = \deg \Psi$.*

2.3. Some counterexamples

2.3.1. Example 2.1. This trivial example shows that the condition of the equiv-ariant homotopy of Φ and Ψ on N in Theorem 2.1 is essential. Let $\beta : S^1 \to S^1$ be a symmetry on S^1 with respect to some diameter $[A, B]$; this defines a Z_2-action on S^1. Consider two maps $\Phi(x) = x$ and $\Psi(x) = A$ on S^1. It is easy to see that Φ and Ψ are equivariant (but not equivariantly homotopic on $\{A, B\}$); however, $\deg \Phi = 1$, $\deg \Psi = 0$.

2.3.2. Example 2.2. The following example shows that the hypothesis that G must be (algebraically) isomorphic to a compact and not completely disconnected group in Corollary 2.9 is essential.

The construction of the example is based on

Lemma 2.4. *Let Γ be the group of all homeomorphisms of the n-dimensional sphere $(n \geq 1)$. There exists a subgroup $\bar{Q}_p$ of the group Γ which is (algebraically) isomorphic to the group Q_p of entire p-adic numbers.*

(It is well-known that the group Q_p in turn is compact and completely disconnected (see, for instance, [Bo1], p. 85).

Proof of Lemma 2.4. Clearly R^+ – the additive group of real numbers – may
be embedded into Γ. We claim that Q_p may be imbedded (in the algebraic sense)
into R^+. Indeed, Q_p is an abelian torsion free group, hence it may be embedded into
a divisible abelian torsion free group Γ_1 (see, e.g. [Ku], p. 165). Clearly, we may
assume that Γ_1 is continual. It remains to show that Γ_1 is isomorphic to R^+. Since
Γ_1 is a torsion free group it may be represented by a continual family of copies of Q,
where Q is an additive group of rational numbers. But R^+ may be also represented
by a continual family of copies of Q and hence Γ_1 is isomorphic to R^+. The lemma
is proved.

Now we give an example of R^+-action (and by Lemma 2.4 of a $\bar{Q}_p$-action) on
the circle, and of two $\bar{Q}_p$-equivariant maps $\Phi, \Psi : S^1 \to S^1$ which are equivariantly
homotopic on $\mathrm{Fix}(\bar{Q}_p, S^1)$, however $\deg \Phi \neq \deg \Psi$. It is more convenient for us to
represent R^+ by R_+ – a multiplicative group of positive real numbers. We consider
S^1 as the interval $[0, 2]$ (identifying 0 and 2) and define for any positive real α a
homeomorphism $f_\alpha : S^1 \to S^1$ by the formula:

$$f_\alpha(x) = \begin{cases} x^\alpha & \text{if } x \in [0,1] \text{ ;} \\ 2 - (2-x)^\alpha & \text{if } x \in [1,2] \text{ .} \end{cases}$$

Denote by $V = \{f_\alpha : \alpha \in R^+\}$ the group of homeomorphisms defined above.
Obviously, V is isomorphic to R_+ and hence contains $\bar{Q}_p$. Now let $\Phi(x) = x$ and
$\Psi(x) = 2 - x$. Since $\mathrm{Fix}(V, S^1) = \mathrm{Fix}(\bar{Q}_p, S^1) = \{0,1\}$, the maps Φ and Ψ are
equivariant and equivariantly homotopic on $\mathrm{Fix}(\bar{Q}_p, S^1)$. However, $\deg \Phi = 1$ and
$\deg \Psi = -1$.

Thus Corollary 2.9 fails for infinite groups which are isomorphic to compact
completely disconnected ones.

On the other hand, the same example with $G = R^+$ shows that this corollary
fails in general for infinite connected groups which are non-compact.

2.3.3. Example 2.3. This example shows that Theorem 2.1 is false for maps into
arbitrary compact manifolds different from spheres. Let $T = S^1 \times S^1$ be the two-
dimensional torus of all points $x = (e^{is}, e^{it})$. Let Z_m be a cyclic group of order
$m = pq$, where $p, q > 1$ are relatively prime. Given a generating element U in Z_m
we define the Z_m-action on T by

$$U^k \left(e^{is}, e^{it} \right) = \left(e^{i(s+2\pi k/p)}, e^{i(t+2\pi k/q)} \right) ,$$

where $k = 1, 2, \ldots, m$. Consider a map $\Phi : T \to T$ defined by

$$\Phi\left(e^{is}, e^{it}\right) = \left(\left(e^{is}\right)^{rp+1}, \left(e^{it}\right)^{zq+1}\right) ,$$

where $r, z \in N$; $r, z \neq 1$. By construction Φ is equivariant with respect to the Z_m-action. By Lemma 1.13 we get

$$\deg \Phi = (rp + 1)(zq + 1) = rzpq + rp + zq + 1 . \tag{2.4}$$

Since p and q are relatively prime, from (2.4) it follows that (generally speaking) $\deg \Phi$ is not equal to one modulo m. On the other hand, the identity map $I : T \to T$ is also equivariant with respect to the (free) Z_m-action. However, $\deg I = 1$.

2.3.4.   Observe that our Corollary 2.2 is in sharp contrast with the well-known example of Conner-Floyd (see [Bre], p. 58) of a Z_m-action ($m = pq$ with p and q relatively prime) on the sphere S^3 with the property that for any $s \in Z$ there exists a map $\Phi : S^3 \to S^3$ which commutes with this action and has degree s.

2.4. Historical and bibliographical notes

2.4.1.   Historically, the first result in degree theory for equivariant maps was the famous Ljusternik-Schnirelman-Borsuk Theorem which states that the degree of an odd map of a finite-dimensional sphere into itself is odd [Bor2,Bor3,LS]. This result has been generalized by S. Eilenberg [Ei] to the simplicial maps which commute with one simplicial free action of a cyclic group Z_p (p is prime) on spheres (cf. Corollary 2.1). The further development of the theory was mostly due to P. Smith and M. Krasnoselskii. A particular case of Theorem 2.1 ($M = S$, G is a cyclic group and the action of G on M is free) was considered for the first time in M. Krasnoselskii's [Kr1] paper. This result follows from P. Smith theory (see, for instance, [Le], p. 371) only when it is additionally required that p is prime and the second action is free or semi-free. Later Y. Israilevich and E. Muhamadiev [IM] eliminated the condition that p be prime; in particular, they defined Smith indices for free and semi-free actions of a cyclic group of arbitrary finite order and proved an analogue of formula (2.3) for arbitrary p.

2.4.2.   A "completely homological" proof of the theorem of Krasnoselskii was obtained by Ju. Borisovich and Y. Izrailevich [BoI] who used the Borel spectral sequence. In [Za1] for the maps which are equivariant with respect to the actions of a cyclic group on spheres ($N = \text{Fix}(G, M)$) Corollary 2.4 was proved (in the framework of the geometric approach) by P. Zabrejko under the additional "regularity" condition. This condition was dropped by Ju. Borisovich, Y. Izrailevich and T. Schelokova [BoISc,Sc1,Sc2,Sc3,Sc4]. Later on a corresponding geometric proof was suggested by P. Zabrejko [Za2]. For arbitrary finite group actions and for maps from a sphere to a sphere, Corollary 2.3 was established by Z. Balanov and S. Brodsky [BB1,BB2] (see also [Kos]); some modification of their result can be found in [Ba1]. For maps from a cohomological sphere to a cohomological sphere the formula which expresses the degree of an equivariant (with respect to finite group actions) map in terms of the so-called equivariance indices was established by T. Fomenko [BF].

Theorem 2.1 was proved by Z. Balanov and A. Kushkuley [BK1, BK2, Ba2, BKZ1, BKZ2, KB].

Corollary 2.9 was obtained by Z. Balanov and S. Brodsky [BB1] (see also [ZK]).

2.4.3.   If S is a $G - ENR$-space Corollary 2.1 follows from [Kom]. If G acts cellularly Corollary 2.1 follows from [Di1,Di2].

If M and S coincide (as G-spaces) then Corollary 2.2 follows from [Lü] (see also [BoIsc,Sc4,BF] where a situation of cohomological spheres was considered).

For the case of finite group free actions on spheres the trick described in the proof of Corollary 2.7 was used by A. Dold [Do2] (see also [Dac]).

Corollary 2.7 was obtained by Z. Balanov and A. Kushkuley [BKZ1, BKZ2, KB, Ba2].

The situation of free actions of a cyclic group on manifolds was studied by J. Izrailevich [Iz], where the most general situation of a couple of free actions of cyclic groups of the divisible orders was considered (see also survey [BF], Theorem 3).

Corollary 2.8 was obtained by Z. Balanov and A. Kushkuley [BK1, BKZ1, BKZ2, Ba2, KB]. If M and S coincide (as G-spaces) then Corollary 2.8 follows from [Lü].

2.4.4.   Under the assumption that M is a smooth G-manifold Corollaries 2.5 and 2.6 were obtained by C. Bowszyc [Bow1] (see also the expository paper [Bow2]).

2.4.5.   Example 2.2 was constructed by Z. Balanov and S. Vinichenko [BV1], Example 2.3 – by Z. Balanov and A. Kushkuley [Ba1].

The bibliographical comments concerning the case of representation spheres will be given in Chapter 3.

For additional bibliographical information we refer the reader to the survey [St].

Chapter 3

DEGREE THEORY FOR EQUIVARIANT MAPS
OF FINITE-DIMENSIONAL MANIFOLDS:
SMOOTH ACTIONS

In the previous chapter we have studied the degree of equivariant
maps in the general case of topological G-manifolds. In this chapter assuming G to be an arbitrary (not necessarily finite) compact Lie group
we study degrees of equivariant maps from a smooth G-manifold into a
G-representation sphere.

The chapter is organized as follows. In the first section using our
results from Chapters 1 and 2 we introduce some integer-valued characteristics connected with the G-actions on M and S. One may consider these
characteristics as the geometric analogs of the so-called equivariance indices introduced by T. Fomenko (Schelokova) in [Sc4] (see also [BF], [Di1],
[Di2]). In terms of these characteristics we formulate our main results
(Theorem 3.1 dealing with finite group actions, and Theorem 3.2 dealing
with arbitrary compact Lie group actions), and present some corollaries for
p-group actions, torus actions, semi-free actions, abelian group actions and
others. Taking an arbitrary smooth manifold N instead of S and assuming
a group G acting on N in such a way that $N^G \neq \emptyset$, we use some straightforward arguments in order to show that most of our results remain valid
in this situation (Corollary 3.6).

The second section is auxiliary and rather technical in nature. Here
we discuss some properties of the cap-product and Thom class useful in
dealing with "bad" orbit types in M – those for which $\dim M^K > \dim S^K$.

The third section consists of two parts. In the first one we generalize
the classical result on the existence of an invariant tubular neighborhood

as follows: let a finite group G act smoothly and effectively on a compact smooth manifold M, and let $\bar{M}$ be a set of all non-principal orbits. Then there exists an invariant neighborhood U of $\bar{M}$ such that the set $U \setminus \bar{M}$ can be equipped with a natural structure of a one-dimensional foliation (Lemma A).

In the second part of the section by means of the standard piecewise linear topology technique we establish some kind of an (equivariant) "general position" result (Lemma B).

Note that Lemmas A and B together with the "elimination" technique based on the usage of the Thom class are main ingredients of our approach. They come together in the fourth section where we prove our main degree result (Theorem 3.1) assuming G to be a finite group.

In the fifth section using the p-group variant of Theorem 3.1 and the density of p-subgroups in a torus, we get degree results assuming G to be a torus (Corollary 3.3). Combining this result with the well-known properties of the Euler characteristic of orbits of compact Lie group actions, and using Theorem 3.1 once again, we extend our result to the case of an arbitrary compact Lie group (Theorem 3.2).

In the sixth section we describe one trick which allows us to extend our results to a situation of equivariant maps from one G-manifold to another G-manifold without assuming the second manifold to be a sphere (Corollary 3.6).

In the seventh section we present some bibliographical comments.

3.1. Statement of results

3.1.1.   Before stating our results we need some preliminary constructions.

Let G be a compact Lie group, and $B \subset A$ closed subgroups of G. Set $W(A, B) = (N(B) \bigcap A)/B$, $N(B, A) = N(B)/(N(B) \bigcap A)$, where $N(H)$ denotes a normalizer of a subgroup H in G. Note that in general, the quotient $N(B, A)$ is not a group but $N(A, A) = N(A)/A = W(A)$ is the usual Weyl group. In addition, if G is abelian then $N(B, A) = G/A$ and $W(A, B) = A/B$. It is clear that $|W(A, B)||N(B, A)| = |W(B)|$ if the right hand side is finite. Below we denote by T a maximal torus of G.

Assume G acts smoothly on compact n-dimensional smooth manifolds M and N, and, in addition, G acts effectively on M. Let $\mathrm{Or}(G, M) = \mathrm{Or}(M) = \{(H_1), (H_2), ..., (H_m)\}$ be the set of all orbit types in M. Let $f : M \to N$ be an equivariant map. In accordance with Remark 1.1, when we speak about the degree of an equivariant map $f : M \to N$, we assume the $W(H)$-actions on M^H and N^H are concordant for all $(H) \in \mathrm{Or}(M)$.

Definition. Let (B) be an orbit type in $\mathrm{Or}(M)$ such that $\dim M^B = \dim N^B$. Denote by $O_*(M, N, B)$ a subset of $\mathrm{Or}(M)$ containing all the orbit types (H) satisfying the condition: $(H) \geq (B)$ and $\dim M^H = \dim N^H$. We also set $O_*(M, N, \{e\}) = O_*(M, N) = O_*$ (here e is the unity of G; recall that we have assumed the G-action on M to be effective). Finally, for any $(A) \in O_*(M, N, B)$ we denote by $O_*(M, N, A, B)$ the set of all the orbit types $(H) \in O_*(M, N, B)$ which lie between (A) and (B).

Let V be an orthogonal representation of G and let $S(V)$ be the unit sphere in V. Choose some $(B) \in O_*(M, S(V))$ and take an arbitrary $(A) \in O_*(M, S(V), B)$. Let C be a connected component of M^A. Take a point $p \in C$ such that $G_p = A$. Let $\sigma_p(A, B)$ be the unit sphere in the plane $N_p(M^A)$ normal to M^A in M^B. Take also the orthogonal complement W to V^A in V^B. It is clear that the group $N(B) \cap A$ as well as $W(A, B)$ act on $\sigma_p(A, B)$ and on $S(A, B) = S(W)$. Let $F_M = \{M_1 \subset M_2 \subset .... \subset M_m = M\}$ and $F_{S(V)} = \{S(V)_1 \subset S(V)_2 \subset ... \subset S(V)_m = S(V)\}$ be G-invariant filtrations defined by (1.1) and (1.2) respectively (see also Subsection 1.3.5).

Definition. Let us define $O_{**}(M, S(V), B)$ as the set of all orbit types $(A) \in O_*(M, S(V), B)$ such that the pair $(\sigma_p(A, B), S(A, B))$ is $((N(B) \cap A), F_M, F_{S(V)})$-extendable in the sense of Subsection 1.3.5. As a matter of convenience we assume $(B) \in O_{**}(M, S(V), B)$. We also set $O_{**}(M, S(V), \{e\}) = O_{**}(M, S(V)) = O_{**}$. Finally, for any $(A) \in O_{**}(M, S(V), B)$ denote by $O_{**}(M, S(V), A, B)$ the set of all orbit types $(H) \in O_{**}(M, S(V), B)$ which lie between (A) and (B).

Observe that:

1) Since the groups $N(B) \cap A$ and $W(A, B)$ act on $\sigma_p(A, B)$ and $S(A, B)$ in the same way one can require in the above definition the $(W(A, B), F_M, F_{S(V)})$-extendability of the pair $(\sigma_p(A, B), S(A, B))$ (with respect to $W(A, B)$-equivariant maps). The only reason of using $N(B) \cap A$ is that $W(A, B)$ is not a subgroup of G in general (see Subsection 1.3.5).

2) It is obvious that for any $(B) \in O_*(M, S(V))$ one has $O_{**}(M, S(V), (B)) \subset O_{**}(M, S(V))$.

3) Some conditions which ensure that $O_*(M, S(V), B) = O_{**}(M, S(V), B)$ for $(B) \in O_*(M, S(V))$ are described in Corollary 1.6. For example, this equality holds if $\dim M^H \leq \dim S(V^H)$ for all $(H) \geq (B)$.

We are now in a position to associate with any $(B) \in O_*(M, S(V))$ and $(A) \in O_{**}(M, S(V), B)$ an integer $\alpha(M, S(V), A, B) = \alpha(A, B)$ as follows.

By definition, if $(A) \in O_{**}(M, S(V), B)$ and $(A) \neq (B)$ then there exists a $W(A, B)$-equivariant map $f_{A,B} : \sigma_p(A, B) \to S(A, B)$ and, moreover, if $(H) \in O_*(M, S(V), A, B)$ then $(H) \in O_{**}(M, S(V), A, B)$, so that there exists a $W(A, H)$-equivariant map $f_{A,H} : \sigma_p(A, H) \to S(A, H)$. As it is clear from the definition of extendable pairs (see Subsection 1.3.5), we can pick up equivariant maps $f_{A,B}$ in such a way that for any orbit type $(H) \in O_*(M, S(V), A, B)$ one has $f_{A,B}|\sigma_p(A, H) = f_{A,H}$.

Definition.   Let $(B) \in O_*(M, S(V))$ and $(A) \in O_{**}(M, S(V), B)$, $(A) \neq (B)$. Fix a map $f_{A,B}$ and a family of maps $f_{A,H}$, $(H) \in O_*(M, S(V), A, B)$, satisfying the conditons described above. Set $\alpha(M, S(V), A, B) = \alpha(A, B) = \deg f_{A,B}$. If $(A) = (B)$ we set $\alpha(M, S(V), B, B) = 1$.

Suppose now that M is as above and G acts also on a (compact, connected, oriented) manifold N. Assume that all the sets N^H, $(H) \in Or(N)$, are connected and orientable. Suppose also that $N^G \neq \emptyset$. In this case we will define maps $f_{A,B,M,N}$ and numbers $\alpha(M, N, A, B)$ in the following way. Let G act trivially on R^1. Take the G-space $N_1 = N \times R^1$. Let $q \in (N_1)^G$. It is clear that the action of G on $V = T_q(N_1)$ satisfies the condition $\dim S(V^H) = \dim N^H$ for any $(H) \in Or(N)$. We will set $f_{A,B} = f_{A,B,M,N} = f_{A,B,M,S(V)}$ and apply the same obvious redefinition to other notions described above.

3.1.2.   *Henceforth* we assume M to be an n-dimensional Riemannian G-manifold, the G-action on M is effective and S is a G-representation sphere. Note that there is no loss of generality if we assume M to be smooth but not Riemannian. Indeed, it is well-known that any compact smooth G-manifold has a G-invariant Riemannian metric if the acting group is compact (see, for instance, [Bre]).

For the sake of simplicity, in what follows we assume that M^H is connected and orientable for all $(H) \in O_{**}(M, N)$ (H is of maximal rank if G is infinite). As will be seen from the proofs the first of these two conditions is used only to simplify notations. The second condition can be also avoided in some cases. To treat the non-orientable case one has to use the (mod 2)-degree theory. The precise statements do not require any new ideas. Therefore, we do not want to go into details and present one simple result which is true without assuming any orientability condition (see Corollary 3.11). Note by the way, that for a torus or p-group $(p > 2)$ action on M all the connected components of fixed point sets are orientable [Bre].

We complete this subsection with two simple results concerning torus actions.

Proposition 3.1. *Let T be a torus acting on M and S. Let p be a prime and $\{G_i\}$ a chain of p-subgroups of T such that $\bigcup G_i$ is dense in T.*

(a) *There exists $i_0 > 0$ such that $M^{G_i} = M^T$ and $S^{G_i} = S^T$ for all $i \geq i_0$;*

(b) *there exists $i_1 > 0$ such that $(T) \in O_{**}(M, S, T)$ iff $(G_i) \in O_{**}(M, S, G_i)$ for all $i \geq i_1$.*

Proof of Proposition 3.1.

(a) Set

$$X = \bigcap_i M^{G_i}.$$

It is clear that $X = M^{G_j}$ for some j, otherwise we have an infinite chain of closed submanifolds in M. Further, if there exists $x \in X \backslash M^T$ then $t(x) \neq x$ for some $t \in T$. Taking a sequence $g_i \in G_i$, $i = 1, 2, \ldots$, such that $\lim g_i = t$ we get $x = \lim g_i(x) = t(x) \neq x$.

To complete the proof of statement (a) one can apply the same arguments to S, choose the corresponding $G_{j'}$ and set $i_0 = \max\{j, j'\}$.

(b) The "only if" part of this statement is obvious. Suppose therefore, that for any $i > i_1$ there exists a G_i-equivariant map $f_i : \sigma \to \tau$ of appropriate orthogonal spheres. Let f be a condensation point of the set $\{f_i\}$. Clearly, f is continuos. For $g \in T$ find $i_2 > i_1$ such that for all $i > i_2$ there exists $g_i \in G_i$ for which $\|g(x) - g_i(x)\| < 1/4i$, $\|g(y) - g_i(y)\| < 1/4i$, $\|f(x) - f_i(x)\| < 1/4i$ and $\|f(g(x)) - f(g_i(x))\| < 1/4i$ for all $x \in \sigma$ and $y \in \tau$. Then for any $x \in \sigma$ we have: $\|f(gx) - gf(x)\| = \|f(gx) - f(g_i(x)) + f(g_i(x)) - f_i(g_i(x)) + g_i(f_i(x)) - g_i(f(x)) + g_i(f(x)) - gf(x)\| \leq 1/4i + 1/4i + 1/4i + 1/4i = 1/i$. That is a T-equivariant map $\sigma \to \tau$ exists iff the sequence of G_i-equivariant maps exists. It is clear from the above, that the same is true for the equivariant extensions of maps defined on any T-invariant subspace (note that a set X is T-invariant if and only if it is G_i-invariant for all i large enough).

3.1.3. We first state our main result in the case G is a finite group.

Theorem 3.1.   *Let G be a finite group , and let $\Phi, \Psi : M \to S$ be a pair of G-equivariant maps. Then there exist integers $b(K)$, $(K) \in O_{**}(M, S)$, such that for any $(H) \in O_*(M, S)$*

$$\deg(\Phi|M^H) - \deg(\Psi|M^H) = \sum_{(K) \in O_{**}(M,S,H)} \alpha(K, H)b(K)|N(H, K)|.$$

Note that using the Möbius function of the partially ordered set $O_*(M, S)$ (as suggested in [Kom]) one can express the numbers $b(K)$ from the last formula in terms of differences $\deg(\Phi|M^L) - \deg(\Psi|M^L)$, $(L) > (K)$, $(L) \in O_*(M, S)$.

Observe also that the meaning of integers $b(K)$ is further clarified by Theorem 5.5 in Chapter 5.

Suppose that under the conditions of Theorem 3.1 $S^G \neq \emptyset$. Set $\Psi : M \to S^G$ to be a trivial equivariant map such that $\Psi(M) = \mathrm{pt} \in S^G$. Then from Theorem 3.1 it follows immediately

Corollary 3.1.   *Under the above assumptions there exist integers $b(K)$, $(K) \in O_{**}(M, S)$, such that for any $(H) \in O_*(M, S)$*

$$\deg(\Phi|M^H) = \sum_{(K) \in O_{**}(M,S,H)} \alpha(K, H)b(K)|N(H, K)|.$$

Applying Theorem 3.1 and Corollary 3.1 to p-group actions we get the following result.

Corollary 3.2. *With the same assumptions as in Theorem 3.1 suppose G is a p-group.*

(a) *If $(G) \notin O_{**}(M, S)$ then $\deg \Phi \equiv \deg \Psi$ (mod p);*

(b) *if $(G) \in O_{**}(M, S)$ then there exists an integer α which modulo p is completely defined by the G-spaces M and S, such that*

$$\deg \Phi - \deg \Psi \equiv \alpha \cdot (\deg(\Phi|M^G) - \deg(\Psi|M^G)) \quad (\mathrm{mod}\ p);$$

in particular (see Corollary 3.1),

$$\deg \Phi \equiv \alpha \cdot \deg(\Phi|M^G) \quad (\mathrm{mod}\ p).$$

3.1.4.   Since an (infinite) torus T contains dense chains of p-subgroups for any prime p, Corollary 3.2 can be applied to T-equivariant maps (for details see Section 3.5).

Corollary 3.3.   Let $G = T$ be an (infinite) torus and let $\Phi, \Psi : M \to S$ be T-equivariant maps.

(a) If $(T) \notin O_{**}(M, S)$ then

$$\deg \Phi = \deg \Psi$$

is uniquely determined by the T-actions on M and S;

(b) if $(T) \in O_{**}(M, T)$ then there exists an integer $\alpha = \alpha(M, S, T)$ depending only on the T-actions on M and S such that

$$\deg \Phi - \deg \Psi = \alpha \cdot (\deg(\Phi|M^T) - \deg(\Psi|M^T));$$

in particular,

$$\deg \Phi = \alpha \cdot \deg(\Phi|M^T).$$

Following the scheme of the proof of Theorem 3.1 (see Section 3.4) one can easily get

Corollary 3.4.   Retaining all other conditions of Theorem 3.1 suppose the group G is not necessarily finite. Let H be a closed subgroup of G such that $|W(K)| < \infty$ for all $(K) \in \mathrm{Or}(M, G)$, $(K) \geq (H)$. Suppose that $\dim M^H = \dim S^H$. Then Theorem 3.1 holds for M^H, i.e. for equivariant maps $\Phi, \Psi : M \to S$ there exist integers $b(K)$, $(K) \in O_{**}(M, S, H)$ such that

$$\deg(\Phi|M^H) - \deg(\Psi|M^H) = \sum_{(K) \in O_{**}(M, S, H)} \alpha(K, H) b(K) |N(H, K)|.$$

Remark 3.1.   Note that the assumption $|W(H)| < \infty$ implies the finiteness of the quotient $N(H, K)$ for all $K > H$. From this it follows that the last formula makes sense. In particular, Corollary 3.4 is valid when H is of maximal rank (see Subsection 1.1.1).

Combining Corollaries 3.3 and 3.4 and Remark 3.1 (for details see Section 3.5) we obtain

Theorem 3.2. *Let G be a compact Lie group, $\Phi, \Psi : M \to S$ be a pair of G-equivariant maps.*

(a) *If G is infinite and $(T) \notin O_{**}(M, S)$ then*

$$\deg \Phi = \deg \Psi.$$

(b) *If $(T) \in O_{**}(M, S)$ then there exist unique integer α depending only on the G-actions on M and S , and integers $b(K) = b(K, \Phi, \Psi)$, $(K) \in O_{**}(M, S, T)$, such that*

$$\deg \Phi - \deg \Psi = \alpha \cdot \sum_{(K) \in O_{**}(M,S,T)} \alpha(K, T) b(K) \chi(G/K)$$

(here $\chi(\cdot)$ means the Euler characteristic).

(c) *If $(H) \in O_*(M, S)$ and H is of maximal rank, then*

$$\deg \Phi | M^H - \deg \Psi | M^H = \sum_{(K) \in O_{**}(M,S,H)} \alpha(K, H) b(K) |N(H, K)|.$$

Note that if $M^T \neq \emptyset$ then there exists $(H) \in \operatorname{Or}(M)$, $H \supset T$, such that $M^T = M^H$. Hence the notation $\alpha(K, T)$ makes sense even if $(T) \notin \operatorname{Or}(M)$.

As a consequence from Theorem 3.2, we get a generalization of Theorem 2.1 as follows:

Corollary 3.5. *With the same notations as in Theorem 3.2, let $\{(H_i)\}_{i=1}^s$ be all the orbit types of the G-action on M such that subgroups H_i have maximal rank and $(H_i) \in O_{**}(M, S)$.*

Then the degree of an equivariant map $f : M \to S$ is uniquely determined modulo $GCD\{\chi(G/H_i)|i = 1, 2, .., s\}$. In particular, if G is infinite and all the stationary subgroups of the G-action on M are of non-maximal rank then $\deg f$ is uniquely determined by the G-actions on M and S.

3.1.5. In the previous subsections we have been interested in degrees of equivariant maps from a manifold into a sphere. In this subsection we suppose a compact Lie group G acts smoothly on smooth (compact, closed, oriented, connected) manifolds M and N with $\dim M = \dim N$. To obtain degree results in this case we have to make the following assumption with respect to the G-manifold N:

$$N^G \neq \emptyset.$$

This assumption allows us to treat a general "non-spherical" case by means of combining the above mentioned "spherical" results with simple G-manifold technique (for details, see Section 3.6). To keep our monograph in reasonable limits we omit some obvious reformulations.

Corollary 3.6.   *Suppose $N^G \neq \emptyset$ and for any $H \in \mathrm{Or}(N, G)$, N^H is connected and oriented. Then Theorems 3.1 and 3.2 are valid for equivariant maps from M to N. For instance, let $\Phi : M \to N$ be an equivariant map.*

(a) If G is infinite and $(T) \in O_{**}(M, N)$ then there exist integers $b(K) = b(K, \Phi)$, $(K) \in O_{**}(M, N, T)$, such that

$$\deg \Phi = \alpha \cdot \sum_{(K) \in O_{**}(M,N,T)} \alpha(K, T) b(K) \chi(G/K),$$

where the numbers α and $\alpha(K, T)$ depend only on the G-actions on M and N;

(b) if $G = T$ is infinite and $(T) \notin O_{**}(M, N)$ then $\deg \Phi = 0$;

(c) if $G = T$ is infinite and $(T) \in O_{**}(M, N)$ then there exists an integer α which is completely defined by the G-spaces M and N , such that

$$\deg \Phi = \alpha \cdot \deg(\Phi | M^G);$$

(d) if G is a p-group then

$$\deg \Phi \equiv \alpha \cdot \deg(\Phi | M^G) \pmod{p};$$

(e) if under the conditions of (a), (c) or (d) $M = N$ (as G-spaces) then $\alpha = 1$.

Assuming a G-action on M to be semi-free we get from Corollary 3.6 the following result.

Corollary 3.7.   *Suppose that under the assumptions of Corollary 3.6 the G-action on M is semi-free. Then for any equivariant map $\Phi : M \to N$ we have:*

(a) if G is finite and $\dim M^G = \dim N^G$ then

$$\deg \Phi \equiv \deg(\Phi | M^G) \cdot \alpha(G, M, N) \pmod{|G|},$$

and if, in addition, the G-action on N is also semi-free then $\alpha(G, M, N)$ is relatively prime to $|G|$;

(b) if G is finite and $\dim M^G \neq \dim N^G$ then

$$\deg \Phi \equiv 0 \pmod{|G|};$$

(c) if G is infinite and $\dim M^G = \dim N^G$ then

$$\deg \Phi = \deg(\Phi|M^G) \cdot \alpha(G, M, N);$$

if, in addition, the action of G on N is also semi-free then $\alpha(G, M, N) = \pm 1$;

(d) if G is infinite and $\dim M^G \neq \dim N^G$ then

$$\deg \Phi = 0.$$

Using Corollary 3.7 one can get some information on relations between the numbers $\alpha(K, H)$ (for details, see Section 3.6).

Corollary 3.8.   Let M^n and N^n be smooth (compact, closed, connected, orientable) G-manifolds. Take orbit types $(L), (K), (H) \in O_{**}(M, N)$ such that $(H) < (K) < (L)$ is a dense chain of orbit types. Then

$$\alpha(L, H) \equiv \alpha(L, K) \cdot \alpha(K, H) \pmod{|W(K, H)|}.$$

3.1.6.   In this subsection we consider abelian group actions.

Note that if G is an abelian group then for any subgroups $H \subset K \subset G$ one has $W(K, H) = (N(H) \cap K)/H = K/H$ and $N(H, K) = N(H)/(N(H) \cap K) = G/K$.

Assuming G to be abelian one can strengthen Corollary 3.8 in the following way (for details, see Section 3.6).

Corollary 3.9.   Suppose that under the conditions of Corollary 3.8 the acting group G is abelian. Then one can pick up $\alpha(L, H)$, $\alpha(L, K)$ and $\alpha(K, H)$ in such a way that $\alpha(L, H) = \alpha(L, K) \cdot \alpha(K, H)$.

Now using Corollary 3.9 we will consider a special case of Theorem 3.2 assuming, in addition, the orbit types in $O_{**}(M, S, T)$

$$\{(H_1), (H_2), ..., (H_s)\}$$

form a linear chain, i. e.

$$(H_1) > (H_2) > ... > (H_s).$$

Take two arbitrary orbit types (H_i) and (H_j) so that $i > j$. Then by Corollary 3.9 one can pick up the numbers $\alpha(H_{i-1}, H_i), \ldots, \alpha(H_j, H_{j+1})$ and $\alpha(H_j, H_i)$ so that

$$\alpha(H_j, H_i) = \alpha(H_j, H_{j+1}) \cdot \alpha(H_{j+1}, H_{j+2}) \cdot .... \cdot \alpha(H_{i-1}, H_i).$$

This speculation yields immediately the following result.

Corollary 3.10.   *Assuming all the conditions of Theorem 3.2 suppose, in addition, that G is abelian, $(T) \in O_{**}(M, S)$ and the orbit types*

$$(H_1) > (H_2) > ... > (H_s)$$

in $O_{**}(M, S, T)$ form a linear chain. Let $\Phi, \Psi : M \to S$ be a pair of G-equivariant maps. Then one can pick up the integers $\alpha(H_j, H_{j+1})$, $j = 1, \ldots, s-1$, and b_j, $j = 1, \ldots, s$, so that for any H_i

$$\deg(\Phi|M^{H_i}) - \deg(\Psi|M^{H_i}) = \sum_{j<i} \alpha(H_j, H_{j+1}) \cdot \ldots \cdot \alpha(H_{i-1}, H_i) \cdot b_j \cdot |G/H_j| + b_i \cdot |G/H_i|$$

where the integers $\alpha(H_j, H_{j+1})$ are unique modulo $|H_j/H_{j+1}|$ and depend only on the G-actions on M and S .

Taking an arbitrary (compact, closed, oriented, connected) n-dimensional Riemannian manifold N instead of S, and assuming $N^G \neq \emptyset$ one can easily reformulate Corollary 3.10 in a manner which is compatible with Corollary 3.6.

The following result can be proved by means of the methods developed in this chapter.

Corollary 3.11.   *Let the group $G = Z_2$ act semi-freely on n-dimensional (compact, closed, oriented, connected) smooth manifolds M and N. Let N^G be connected and let $\{M_i\}$ be connected components of M^G. Then for any equivariant map $f : M \to N$*

$$\deg f \equiv \sum_{i:\ \dim M_i = \dim N^G} \alpha_i \cdot \deg(f|M_i) \pmod{2}$$

where α_i depend only on the Z_2-actions on M and N and are uniquely defined modulo 2 (here $\deg f$ denotes the usual Z-degree if both M and N are orientable, and it is the Z_2-degree if at least one of them is non-orientable).

It should be noticed that we do not assume any orientability conditions with respect to M_i's.

3.1.7.   We complete this section with the "relative form" of Theorem 3.2.

Suppose that under the conditions of Theorem 3.2 $A \subset M$ is a closed invariant subset. The following theorem follows directly from the proof of Theorem 3.2 (see also Corollary 3.4 and Remark 3.1).

Theorem 3.2'.   Let $\Phi, \Psi : M \to S$ be a pair of equivariant maps such that Φ is equivariantly homotopic to Ψ over A.

(a) Let G be infinite. If $(T) \notin O_{**}(M \backslash A, S)$ or if $M^T \subset A$ then $\deg \Phi = \deg \Psi$;

(b) if $(T) \in O_{**}(M \backslash A, S)$ then there exist a unique integer α, which depends only on the actions of G on M and S , and integers $b(K) = b(K, \Phi, \Psi)$, $(K) \in O_{**}(M \backslash A, S, T)$, such that

$$\deg(\Phi) - \deg(\Psi) = \alpha \cdot \sum_{(K) \in O_{**}(M \backslash A, S, T)} \alpha(K, T) b(K) \chi(G/K);$$

(c) if $(H) \in O_*(M \backslash A, S)$ is of maximal rank, then

$$\deg(\Phi | M^H) - \deg(\Psi | M^H) = \sum_{(K) \subset O_{**}(M \backslash A, S, H)} \alpha(K, H) b(K) |N(H, K)|.$$

Note that all the results presented in this section can be stated in the same relative form.

Other related results will be discussed in Chapter 5.

3.2. Thom class and cap-product

3.2.1.   In this section we recall shortly some standard properties of the $\bigcap$-product and Thom class. We follow [Do1] where the reader can find all the necessary details. In what follows, unless explicitly stated otherwise, we will use integer (co)homology groups.

Let (X, A_1, A_2) be an excisive triad (e. g. A_1 and A_2 are open in X). By the definition of the $\bigcap$-product (cf. e. g. [Do1]) we have

$$\bigcap : H^{n-k}(X, A_2) \bigotimes H_n(X, A_1 \bigcup A_2) \to H_k(X, A_1).$$

If (Y, B_1, B_2) is another excisive triad and if $f : (X, A_1, A_2) \to (Y, B_1, B_2)$ is a map of triads then we have

Lemma 3.1 (see [Do1]). For any $x \in H^{n-k}(Y, B_2)$ and $\xi \in H_n(X, A_1 \bigcup A_2)$

$$f_*((f^*x) \bigcap \xi) = x \bigcap (f_*\xi).$$

Recall now the notion of the Thom class. Let M^{n+k} be an $(n+k)$-dimensional oriented manifold and let N be an n-dimensional submanifold. Then $H_k(M, M\backslash N)$ is freely generated by elements ν_λ which correspond to the components N_λ of N (see [Do1], p. 315). The class ν_λ is said to be a transverse class of N_λ in M. Further, for every component N_λ of N choose an element $g_\lambda \in Z$; then (see [Do1], p. 316) there exists a unique class $y \in H^k(M, M\backslash N)$ such that $< y, \nu_\lambda > = g_\lambda$ for every λ (here ν_λ is a transverse class). In particular, there exists a unique class $\tau_N^M \in H^k(M, M\backslash N)$ such that $\tau_N^M(\nu_\lambda) = 1$ for every λ; it is called the *Thom class* of N in M.

As is well-known (see, for instance, [Do1], p. 317), the Thom class is natural with respect to inclusions $i : U \to M$ of open subsets, i. e. $i^*(\tau_N^M) = \tau_{U \cap N}^U$.

Lemma 3.2 (see [Do1], p. 319). *Suppose M and N are as above, $X \subset N$ is compact, $W \subset M$ is open and $(N\backslash X) \subset W \subset (M\backslash X)$. Let $i : (N, N\backslash X) \subset (M, W)$ be an inclusion. Then*

$$i_*(O_X^N) = \tau_N^M \bigcap O_X^M.$$

3.2.2. Let S_1 and S_2 be two copies of the n-dimensional sphere. Suppose G is a finite group acting orthogonally and semi-freely on both spheres. If $\dim S_1^G \neq \dim S_2^G$ then for any pair of equivariant maps $\Phi, \Psi : S_1 \to S_2$

$$\deg \Phi \equiv \deg \Psi \quad (\text{mod } |G|) \tag{3.1}$$

(see, for instance, [BKZ2]). Following the scheme of the proof of Theorem 2.1 one can compare Φ with the map $\Psi \equiv \text{pt} \in S_2^G$. This leads to the following interpretation of (3.1): the (local) degree of an extension F around the set of zeros contained in $S_1^G \times [0, 1]$ is equal to zero.

Below we study the connection between the (local) degree of a map and the corresponding Thom class. In particular, this connection will allow us to treat the above mentioned "dimensional effects" for arbitrary (not necessarily semi-free) compact Lie group actions.

The following lemma clarifies our approach.

Let U be an oriented n-dimensional manifold, V an n-dimensional vector space and D its k-dimensional subspace ($k > 0$). Assume W_1 is an open subset of V such that W_1 is contractible to $D \backslash \{0\}$ and $W_1 \bigcup (V \backslash D) = V \backslash \{0\}$. Let $f : U \to V$ be a proper continuous map and $K = f^{-1}(0)$. Suppose, finally, that there exists an open subset $\hat{U} \subset f^{-1}(W_1)$ such that $\hat{U} \bigcup (U \backslash f^{-1}(D)) = U \backslash K$.

Lemma 3.3.   If $f^*(\tau_D^V) = 0$ then $\deg_0 f = 0$.

   Proof. Let O_0^V and O_K^U be fundamental classes of V and U at 0 and K respectively. By our assumptions f is a map of triads

$$(U, \hat{U}, U \setminus f^{-1}(D)) \to (V, W_1, V \setminus D).$$

By Lemma 3.2 and our assumptions $\tau_D^V \cap O_0^V = i_*(O_0^D) \in H_k(V, W_1) \cong H_k(D, D \setminus \{0\})$ is non-trivial. Hence, if $\deg_0 f \neq 0$ then using Lemma 3.1 we have

$$f_*(f^*(\tau_D^V) \cap O_K^U) = \tau_D^V \cap f_*(O_K^U) = \tau_D^V \cap (\deg_0 f) \cdot O_0^V =$$

$$(\deg_0 f) \cdot (\tau_D^V \cap O_0^V) = (\deg_0 f) \cdot i_*(O_0^D) \neq 0$$

from which it follows that $f^*(\tau_D^V)$ must be non-trivial.

   The lemma is proved.

   To take advantage of this lemma for proving $\deg_0 f = 0$ one should provide $f^*(\tau_D^V) = 0$. The simplest result in this direction can be formulated as follows.

Lemma 3.4.   If under the assumptions of Lemma 3.3 $H^{n-k}(U, U \setminus f^{-1}(D)) = 0$ then $\deg_0 f = 0$.

   Consider a more complicated situation.

   Let $N^l \subset M^n$ be compact manifolds and $U \subset M^n$ an open subset such that $N^l \cap U$ is connected. Let also V be an n-dimensional vector space and let $D \subset V$ be a k-dimensional subspace $(k > 0)$.

Lemma 3.5.   Under the above assumptions assume that $f : U \to V$ is a continuous map $(K = f^{-1}(0))$ satisfying the following conditions:

   (a) there exist compact submanifolds $N_1, N_2, ..., N_s$ in M and linear subspaces $D_1, D_2, ..., D_s$ in V such that $N_i \supset N = N_0$, $D_i \supset D = D_0$, N_i is not contained in U and $f(N_i \cap U) \subset D_i$ for all $i = 0, 1, 2, ..., s$;

   (b) the set

$$E = \bigcup_{\dim N_j > \dim D_j} N_j$$

is nonempty and $f^{-1}(D) \subset N \cup E$;

   (c) if $\dim N_j > \dim D_j$ for some $0 \leq j \leq s$ then for some linear subspace $P \supset D_j$ $(P \neq D_j)$ one has $f(x) \in P$ only if $f(x) \in D_j$;

   (d) f is a proper map.

   Then:

a) if $l > k$ then $\deg_0 f = 0$;

b) if $l = k$ and $\deg_0 f \neq 0$ then $N \bigcap K \neq \emptyset$ and the restriction of f onto some disk $\sigma \subset U$ orthogonal to N and centered at $x \in N \bigcap K$ maps $\partial \sigma$ into $V \setminus D$.

3.2.3. Proof of Lemma 3.5.

Note that under our assumptions there exist open subsets W_1 of V and $\hat{U} \subset f^{-1}(W_1)$ satisfying the conditions of Lemma 3.3. Hence, f induces a map of triads:

$$(U, \hat{U}, U \setminus f^{-1}(D)) \to (V, W_1, V \setminus D).$$

a) If $\dim N = l > k = \dim D$ then by condition (c) $\deg_0 f = 0$.

Therefore, we assume that

b) $l = k$.

Let $D^\perp$ be the orthogonal complement to D in V and let $p : V \to D^\perp$ be the orthogonal projection of V onto $D^\perp$. Consider the exact sequence of the triple $(U, U \setminus N, U \setminus f^{-1}(D)))$

$$\to H^{n-k}(U, U \setminus N) \xrightarrow{j^*} H^{n-k}(U, U \setminus f^{-1}(D)) \xrightarrow{i^*}$$

$$\to H^{n-k}(U \setminus N, U \setminus f^{-1}(D)) \to \qquad (3.2)$$

Since f induces the map of triads, $f^*(\tau_D^V) \in H^{n-k}(U, U \setminus f^{-1}(D))$. Let us prove, first of all, that under our assumptions $f^*(\tau_D^V) \in \operatorname{Im} j^*$. By the exactness of (3.2) it suffices to prove that the composition $i^* \circ f^*$ is trivial. We argue indirectly and suppose this composition is non-trivial. Then we can assume that there exists a closed set $R \subset U \setminus N$ such that the restriction of f onto $(R, R \setminus f^{-1}(D))$ is cohomologically non-trivial. Let N_t be the minimal submanifold among N_j's such that $\dim N_j > \dim D_j$. Using an excision we can assume that R lies in an arbitrarily small neighborhood of $f^{-1}(D)$ and hence R does not intersect any of $N_j \notin E$. Applying condition (c) to N_t and to the map $p \circ f$ we get a contradiction with the cohomological non-triviality of $f|(R, R \setminus f^{-1}(D))$.

Therefore, the composition $i^* \circ f^*$ is trivial, and hence $f^*(\tau_D^V)$ belongs to the image of j^*.

Let us assume that $\deg_0 f \neq 0$. Hence (Lemma 3.3) $f^*(\tau_D^V) \neq 0$. Combining this with $f^*(\tau_D^V) \in \operatorname{Im} j^*$ and using the definition of the Thom class it is easy to see that:

1) $N \bigcap K \neq \emptyset$;

2) for some disc $\sigma \in U$ orthogonal to N and centered at $x \in N \bigcap K$ one has $0 \notin p \circ f(\partial \sigma)$. Hence $f(\partial \sigma) \bigcap D = \emptyset$.

Lemma 3.5 is proved.

3.3. Invariant foliations and equivariant transversality

Our approach to the degree results stated in Section 3.1 is essentially based on two lemmas presented in this section (see Lemmas A and B).

3.3.1. Before stating Lemma A (generalizing the theorem on the existence of an invariant tubular neighborhood) we need some preliminaries.

Lemma 3.6. *Let a finite group G act effectively on a smooth compact Riemannian manifold M. Let M' be a union of all non-principal orbits of the action of G on M. There exists a closed invariant neighborhood $U \supset M'$, a smooth invariant function $f : U \to R$ and a number $c > 0$ such that:*

(a) f is non-negative on U and $f^{-1}(0) = M'$;

(b) $\mathrm{grad} f \neq 0$ in $f^{-1}((0, c])$.

Proof. First of all, note that since G is finite and acts effectively,

$$M' = \bigcup_{(H) \in \mathrm{Or}(M),\; H \neq e} M^H.$$

Let $O_0(M)$ be the set of minimal orbit types, i. e. $(H) \in O_0(M)$ if and only if $H \neq e$ and M^H is not contained in any M^K with $(K) \in \mathrm{Or}(M)$, $(K) \neq (H)$. Set $d_1 = \min\{\mathrm{dist}(M^H, M^K),\ (H), (K) \in O_0(M),\ M^H \cap M^K = \emptyset\}$. The function $F_H(x) = \mathrm{dist}(x, M^H)$, $(H) \in O_0(M)$, is smooth for all x sufficiently close to a manifold M^H. Let $1 > d_2 > 0$ be such that F_H is smooth inside $(F_H)^{-1}([0, d_2])$ for all H, $(H) \in O_0(M)$. Choose $0 < d < \min\{d_1, d_2\}$ (in what follows we choose d more carefully in accordance with certain additional requirements (see Lemma 3.7(iii))). Take a smooth non-negative function $w : R_+^1 \to R_+^1$ such that w is the identity function from 0 to $d/3$, increases monotonicly from $d/3$ to 1 in $[d/3, d/2]$ and $w(r) = 1$ for $r \geq d/2$ (hence $w'(r) > 0$ for $0 < r < d/2$). It is clear, that the function $r_{d,H} = w \circ F_H$ is smooth everywhere on M and that $\mathrm{dist}(x, M^H) < d/2$ if $r_{d,H}(x) \neq 1$. Define a function $r_d : M \to R^1$ in the following way:

$$r_d = \prod_{(H) \in O_0(M)} r_{d,H}.$$

We list some obvious properties of this function in the following

Lemma 3.7.

(i) *The function r_d is invariant;*

(ii) $(r_d)^{-1}(0) = M'$;

(iii) *For any $\varepsilon > 0$ there exists $d > 0$ such that for any $x \in M$ with $0 < r_d(x) < 1$*
one has:

$$L_d = \bigcap_{r_{d,H}(x) \neq 1} M^H \neq \emptyset \qquad \text{and} \qquad \text{dist}(x, L_d) < \varepsilon;$$

(iv)

$$\text{grad}_x(r_d) = r_d(x) \cdot \sum_{r_{d,H}(x) \neq 1} \frac{w'(F_H(x))}{r_{d,H}(x)} \cdot \text{grad}_x(F_H).$$

Continuation of the proof of Lemma 3.6. Take d sufficiently small to provide
the following property (see Lemma 3.7(iii)): for each $u \in M$ with $0 < r_d(u) < 1$ one
has $L_d \neq \emptyset$.

Choose $x \in M$ with $0 < r_d(x) < 1$. Let γ be a geodesic which starts at x in
the direction of $\text{grad}_x(F_H)$ and let $y = \gamma \cap M^H$. Under our assumptions on d the
length of the geodesic segment $[x, y]$ is equal to $F_H(x)$. In the same way take $z \in L_d$
such that the length l of the geodesic segment $[x, z]$ is equal to $\text{dist}(x, L_d)$. In the
tangent space $T_x(M)$ we have two vectors : $\text{grad}_x(F_H)$ and the velocity v of the
geodesic $[x, z]$. Let α_H be the angle between these two vectors. From the geodesic
triangle $\tau_H = \Delta xyz$ we have $\cos \alpha_H = F_H(x)/l + o(\tau_H)$, where $|o(\tau_H)| \cdot l/F_H(x)$ is
small for small l. In particular, using the compactness of M one can find $\varepsilon_0 > 0$
such that for any $x \in M$ with $\text{dist}(x, L_d) < \varepsilon_0$ one has:

$$\cos \alpha_H = \frac{F_H(x)}{\text{dist}(x, L_d)} + \delta_H(x) \tag{3.3}$$

where

$$|\delta_H(x)| < \frac{1}{2} \cdot \frac{F_H(x)}{\text{dist}(x, L_d)}. \tag{3.4}$$

Using Lemma 3.7(iii) once again decrease d so that for all $x \in M$ with $0 < r_d(x) < 1$
one has $\text{dist}(x, L_d) < \varepsilon_0$.

Now with the last d in hands let us compute a projection of $\text{grad}_x(r_d)$ onto v if
$0 < r_d(x) < 1$. Bearing in mind that $\|\text{grad}_x(F_H)\| = 1$ we get from Lemma 3.7(iv)

$$\frac{< v, \text{grad}_x(r_d(x)) >}{r_d(x)} = \sum_{r_{d,H}(x) \neq 1} \left(\frac{w'(F_H(x)) \cdot F_H(x)}{l \cdot r_{d,H}(x)} + \frac{\delta_H(x) \cdot w'(F_H(x))}{r_{d,H}(x)} \right). \tag{3.5}$$

From (3.3) and (3.4) it follows that all the summands in (3.5) are positive. This
shows that the $\text{grad}_x(r_d)$ is positive for sufficiently small d and all x such that
$0 < r_d(x) < 1$.

To complete the proof of Lemma 3.6 take $r_d : M \to R$ with a sufficiently small d, choose $0 < c < 1$ and set $U = r_d^{-1}([0, c])$, $f = r_d | U$.

As an easy consequence from Lemma 3.6 we have

Lemma A (on a generalized invariant tubular neighborhood).   Let a finite group G act effectively on a smooth compact manifold M. Let M' be a union of all non-principal orbits of the action of G on M. There exists a closed invariant neighborhood U of M' such that:

(a) ∂U is a smooth submanifold;

(b) there exists a parametrized family of smooth curves $\gamma(x, t)$, $0 \leq t \leq 1$, $x \in \partial U$, such that $\gamma(x, 0) \in M'$, $\gamma(x, 1) = x \in \partial U$ and $\gamma(x, t)$, $0 < t \leq 1$ is an invariant one-dimensional foliation of $U \setminus M'$.

Proof. Take a gradient flow defined by a function which was constructed in Lemma 3.6.

Remark 3.2.   Under the notations of Lemma A, define a map $\bar{\pi} : U \to M'$ as follows. Take $y \in U$. If $y \in M'$ we set $\bar{\pi}(y) = y$. Let $y \in U \setminus M'$. Take the point $x \in \partial U$ such that $y = \gamma(x, t_0)$ for some $t_0 \in (0, 1]$, and set $\bar{\pi}(y) = \gamma(x, 0)$. Set also $\pi = \bar{\pi} | \partial U$. It is clear, that the map $\pi | \pi^{-1}(M^K) : \pi^{-1}(M^K) \to M^K$, $(K) \in \mathrm{Or}(M)$, $K \neq e$, is smooth, and hence the map π does not increase dimension.

Remark 3.3.   We will apply Lemma A to the $W(H)$-action on a manifold M^H and we will use the set $M^{>H} = \bigcup M^K$, $K > H$, $(K) \in \mathrm{Or}(M)$, instead of M'. It is clear that Lemma 3.6 (and hence Lemma A) is valid in this situation. In this case we will denote the maps described in Remark 3.2 by $\bar{\pi}_H$ (relatively, by π_H).

Remark 3.4.   It is easy to see that Lemma 3.6 as well as Lemma A are valid for arbitrary compact Lie groups (everything one has to do is to replace a function $r_{d,H}$ by the integral over G/H). Moreover, following the scheme described above one can easily reformulate and prove Lemma A in a non-equivariant fashion (for a finite family of submanifolds). We omit details here.

3.3.2. The following lemma can be deduced from the results on "general position" of piecewise linear maps presented in [Zee] (see also [Mc]). For completeness we present simple straightforward arguments which are sufficient for our purposes.

Lemma B (cf. [Zee, Mc]). *Let V be an orthogonal $(d+1)$-dimensional representation of a finite group G and B^{d+1} a unit ball in V. Let G act freely on a compact $(d-k)$-dimensional manifold X $(k \geq 1)$. For any finite set of linear subspaces $L_j \subset V$, $j = 1, \ldots, m$, there exists an equivariant map f from X to B^{d+1} such that $\dim f^{-1}(G(B^{d+1} \bigcap L_j)) \leq \dim L_j - k - 1$ for all $j = 1, 2, \ldots, m$, provided $\dim L_j \geq k$.*

Proof. By Lemma 1.18 we can assume that the action of G on X is simplicial. Let T_i denote an i-dimensional skeleton of X. Set $P_j = B^{d+1} \bigcap L_j$ and $r_j = \dim P_j$. Without loss of generality one can assume that $r_1 \geq r_2 \geq \ldots \geq r_m$. Since $B \setminus G(\bigcup_j P_j)$ is $(d - r_1 - 1)$-connected , there exists an equivariant map f_{d-r_1} : $T_{d-r_1} \to B^{d+1} \setminus G(\bigcup_j P_j)$. Assume by induction that for all $i \leq r_1$, $(i \geq 2)$ there exists an equivariant map $f_{d-i} : T_{d-i} \to B^{d+1}$ such that $f(T_{d-i})$ is a simplicial complex of dimension $d - i$ and that $\dim f_{d-i}^{-1}(G(P_j)) \leq (r_j - i - 1)$. Let D_{d-i+1} be a (simplicial) fundamental domain in T_{d-i+1}. Take a simplex $\sigma \in D_{d-i+1}$ and let $Q_j = f_{d-i}^{-1}(G(P_j)) \bigcap \partial \sigma$. Let J_i be the join of $f(\partial \sigma)$ with all L_j's such that $\dim L_j < d - i$. Clearly, $\dim J_i \leq d$. Extend f_{d-i} over σ in the following way. Take a barycenter p of σ , set $f_{d-i+1}(p) = 0$ and for any other point $x \in \sigma$ define

$$f_{d-i+1}(x) = \frac{\operatorname{dist}(p, x)}{\operatorname{dist}(p, z)} \cdot f_{d-i}(z),$$

where $z \in \partial \sigma$ is a "central projection" of x with respect to p. In other words, σ is mapped onto the cone over $f(\partial \sigma)$ with the vertex at the origin O of V. Note also, that $f_{d-i+1}^{-1}(P_j) \bigcap \partial \sigma$ is a cone over Q_j. It is clear from this construction that if $r_j + i \geq d$ then $\dim(f_{d-i+1}^{-1}(P_j)) = \dim(f_{d-i}^{-1}(P_j)) + 1 \leq r_j - i$. To make sure that $f(\sigma) \bigcap P_k = \emptyset$ if $\dim L_k < d - i$, we can replace the point O with a point $O_1 \notin J_i$ such that $\dim(f_{d-i+1}^{-1}(P_j))$ will not increase if we shift the vertex of our cone to O_1 (cf. [Zee]). To complete the induction step one has to apply the described procedure to all simplexes from D_{d-i+1} and to use Lemma 1.16.

3.4. The case of a finite group

3.4.1.   The proof of Theorem 3.1 follows the scheme of the proof of Theorem 2.1 (see also Theorem 2.1'). During this section we keep for C, C_i, B, B_i, K, f_0 and F the same notations as in the proof of Theorem 2.1' (see Subsections 2.1.1 and 2.1.2). Assuming the set $Or(M)$ to be equipped with the standard partial order we will use the induction over the orbit types for extending f_0 to F. At the same time the additional requirements on the G-actions on M and S allow us to extend f_0 to F in a more delicate way. Namely, our goal is to construct an extension F for which $F_*(O_K)$ is more "computable" than for the one in the proof of Theorem 2.1' (compare the formulation of Lemma 2.2 with the one of Lemma 3.8).

Recall that the main idea for making an induction step in the proof of Theorem 2.1' was to "build a wall" which separates zeros obtained on the previous steps. Since we had not assumed any smoothness conditions with respect to the G-manifold M we constructed these "walls" from boundaries of more or less arbitrary neighborhoods. The smoothness conditions we have assumed in this chapter allow us to "polish the walls".

Note also that the "walls" must be built only in $O_*(M, S)$. The obvious candidate for a "wall" is a boundary of generalized tubular neighborhoods provided by Lemma A. Moreover, Lemma B gives us a collection of "prefabricated walls" with equivariant maps already defined on them. By means of these maps we can construct F in such a way that all the stuff outside $O_*(M, S)$ can be eliminated by Lemma 3.5. More precisely, we construct our F by extending the maps on the "walls" inside generalized tubular neighborhoods along the geodesics which form the foliations provided by Lemma A. Hence by Lemma B we know subsets which will be mapped in such a way that will prevent us from using Lemma 3.5 (these subsets are described in Subsection 3.4.3). Therefore, we construct our F by induction in such a way that it will not have zeros on the subsets in question since they have nice dimensions (see (3.6)).

3.4.2.   In this subsection we give the precise formulation of the properties which are required from the extension F. Assuming the existence of such an extension we deduce Theorem 3.1.

We want to construct an equivariant map

$$F : C = M \times I \to B$$

such that the following conditions are fulfilled:

(i) For any $(H) \in O_*(M, S)$ there exists a $W(H)$-invariant neighborhood U of $C^{>H}$ in C^H such that $\Gamma = (F|U)^{-1}(0) \subset C^{>H}$.

(ii) Let $(H) \in O_*(M, S)$. Let Γ be as in (i) and let Z be a connected component of Γ. Let (L) be the minimal orbit type (with respect to the partial order on $\mathrm{Or}(M, G)$) among all (R)'s with $Z \cap C^R \neq \emptyset$. Let us take all the orbit types $(L) = (K_0), (K_1), \ldots, (K_s) = (H)$ of $\mathrm{Or}(M, G)$ such that $(L) \geq (K_i) \geq (H)$ and $\dim C^{K_i} \geq \dim B^{K_i}$, $i = 0, 1, \ldots, s$. Then there exists an open neighborhood $U_Z \subset C^H$ of Z such that the map $F|U_Z$, the manifolds $C^L = C^{K_0}$, $C^{K_i} \subset C^H$, and disks $B^L = B^{K_0}$, $B^{K_i} \subset B^H$, $(L) \geq (K_i) \geq (H)$, $i = 0, 1, 2, \ldots, s$, satisfy the conditions of Lemma 3.5.

(iii) If $(H) \in O_*(M, S)$ and $(H) < (K) \in O_{**}(M, S, H)$ then the set $A = (F|(C^K \setminus C^{>K}))^{-1}(0)$ consists of a finite number of isolated points. For any $z \in A$ there exists a trivial $W(K, H)$-invariant neighborhood $U_{z, C^H} = U_{z, C^K} \times D^l$, where the disc D^l ($l = \dim C^H - \dim C^K$) is orthogonal to C^K, and such that the map $F|U_{z, C^H}$ splits into the Cartesian product of its restrictions onto U_{z, C^K} and D^l. Moreover, $F|\partial D^l$ is a $W(K, H)$-equivariant map of degree $\alpha(K, H)$ (see Subsection 3.1.1).

(iv) If $(H), (K) \in O_*(M, S)$, $(H) < (K)$ but $(K) \notin O_{**}(M, S, H)$ then either the map $(F|(C^K \setminus C^{>K}))$ has the same properties as in (iii) above or for any disk D^l orthogonal to C^K in C^H and such that $D^l \cap (F|U)^{-1}(0) \neq \emptyset$, $F(\partial D^l) \cap B^K \neq \emptyset$.

(v) For any $x \in C$ such that $\dim C^{G_x} < \dim B^{G_x}$, $F(x) \neq 0$.

(vi) Let $\{C_i\}$ and $\{B_i\}$ be the filtrations constructed in the proof of Theorem 2.1' (see Subsections 2.1.1 and 2.1.2). For any G-equivariant map $F_i : C_i \to B_i$ satisfying conditions (i)-(v) there exists a G-equivariant extension $F_{i+1} : C_{i+1} \to B_{i+1}$ satisfying the same conditions. That is the map F satisfying (i)-(v) can be constructed by induction over the orbit types and each step of the induction can be carried over regardless of the map obtained on the previous step.

Assume for a moment that such a map exists. Fix some $(H) \in O_*(M, S)$. Following the scheme of the proof of Theorem 2.1' we want to count (local) degrees of "zeros" of the map $f = F|C^H$ and to apply Lemma 2.1.

Take a connected component Z of the set $f^{-1}(0)$. By property (v) for any $z \in Z$ one has $\dim M^{G_z} \geq \dim S^{G_z}$.

Let (L) be the minimal orbit type such that $C^L \cap Z \neq \emptyset$. Consider the following 3 cases.

Case 1. $(L) \notin O_*(M, S, H)$. In this case using property (ii) we can find a neighborhood $U_Z \subset C^H$ of Z such that $\deg f|U_Z = 0$ by Lemma 3.5(a).

Case 2. $(L) \in O_{**}(M, S, H)$. By condition (iii) Z is an isolated point z which has a trivial $W(L, H)$-invariant neighborhood $U_{z,C^H} = U_{z,C^L} \times D^l$ such that the restriction of the map F to this neighborhood satisfies the conditions described in (iii). Hence $\deg(f|U_{z,C^H}) = \deg(f|U_{z,C^L}) \cdot \alpha(L, H)$. Let us count these numbers along the $W(H)$-orbits of singular points. Let $\omega_k = W(H)(z_k)$, $k = 1, 2, ..., s$, be all such orbits in $W(H)(C^L \setminus C^{>L})$. Let V_k be an open neighborhood of ω_k, $k = 1, 2, ..., s$, and $V(L) = \bigcup V_k$. We can assume that the neighborhoods V_k are disjoint and V does not contain any other zeros of f outside $\omega = \bigcup \omega_k$. Bearing in mind Remark 1.1(b) we get

$$\deg f|V_k = |N(H)/(L \cap N(H))| \cdot \deg(f|U_{z_k,C^H}),$$

and therefore

$$\deg(f|V(L)) = |N(H, L)| \cdot \alpha(L, H) \cdot b(L),$$

where

$$b(L) = \sum_k \deg(f|U_{z_k,C^L}).$$

Case 3. $(L) \in O_*(M, S, H) \setminus O_{**}(M, S, H)$. In this case, using conditions (ii) – (iv) and Lemma 3.5(b) we arrive at the following alternative: either $\deg f|U_Z = 0$ as in Case 1 or the situation is the same as in Case 2. Assume that the latter is the case. Then the above (L) is said to be "spurious" orbit type.

Before we start to deal with "spurious" orbit types let us observe that from our considerations in Cases 1 and 2 it follows:

$$\deg \Phi|M^H - \deg \Psi|M^H = \beta_1 + \beta_2 \qquad (*)$$

where

$$\beta_1 = \sum_{(K) \in O_{**}(M,S,H)} \alpha(K, H) \cdot b(K) \cdot |N(H, K)|$$

and

$$\beta_2 = \sum_{(L) \in O_{sp}(M,S,H)} \deg \mu(L, H) \cdot b(K) \cdot |N(H, L)|$$

(here $O_{sp}(M, S, H)$ denotes the set of spurious orbit type occuring in $O_*(M, S, H)$ and $\mu(L, H)$ is the restriction of f onto a sphere $\sigma(L, H) \subset V(L)$ orthogonal to C^L in C^H. Clearly, by condition (iii) $\mu(L, H)$ can be regarded as a map into a sphere orthogonal to S^L in S^H).

We will eliminate the term β_2 using the following strategy:

1) we proceed by induction over the number of the orbit types in $O_{sp}(M, S, H)$ starting with such (L) for which there does not exist $(L)' \in O_{sp}(M, S, H)$ with $(H) < (L)' < (L)$;

2) for such (L) we show that $\deg \mu(L, H) \cdot b(K) \cdot |N(H, L)|$ can be written as the sum

$$\sum_{(H)<(K)<(L),\ (K)\in O_{**}(M,S,H)} \alpha(K, H) \cdot b'(K) \cdot |N(H, K)|;$$

3) we replace in $(*)$ the term $\deg \mu(L, H) \cdot b(L) \cdot |N(H, L)|$ by the above expression and rewrite $(*)$ setting

$$\text{"new } b(K)\text{"} = \text{"old } b(K)\text{"} + b'(K) \qquad (**)$$

for all $(K) \in O_{**}(M, S, H)$ with $(H) < (K) < (L)$;

4) we take the "next" orbit type in $O_{sp}(M, S, H)$ and apply to it 1) - 3).

The realization of the above program gives rise to the following question: how do 1) - 4) affect the general formulae stated in Theorem 3.1, namely:

a) is it true that substitution $(**)$ is "universal", that is, independent of H?

b) how do different induction steps affect each other?

The answer to a) is that whenever $b(K)$, $(K) < (L)$, participates in a formula, $b(L)$ also does, and hence rewriting works "universally".

The answer to b) is hidden in (i) (see also (vi)). Namely, by (i) all zeros originating in different C^K's are isolated from each other. To get 2) (see below) we will work inside a small neighborhood of zeros originating in $C^L \setminus C^{>L}$. Hence our first rewriting does not affect the second one.

Summing up the above arguments it remains to take a spurious (L), to assume 1) and to show 2).

We have in notations of Case 2:

$$\deg(f|V(L)) = \deg \mu(L, H) \cdot b(L) \cdot |N(H, L)|.$$

Note, that we also have

$$\deg(f|W) = \deg \mu(L, H) \cdot b(L) \cdot |N(H, L)|$$

for any sufficiently small smooth manifold with boundary $W \subset C^H$ which contains all the isolated zeros of f in $C^L \setminus C^{>L}$ and does not contain any other singular points of f.

Let i be the minimal integer such that C_i contains C^L. Take F_i (see condition (vi)) and begin the new construction of F from this point. This time since (L) is spurious take a "route" which will lead to a situation when the map on orthogonal sphere to C^L in C^H does not exist. Denote the new F by $\hat{f}$. The zeros of $\hat{f}$ which originate in C^L will now "spill over" into C^H, however, by (i) we will have a $W(H)$-invariant neighborhood of them which does not contain zeros originating in other M^K. We take our W (see above) inside this neighborhood and in such a way that all zeros of $\hat{f}$ originated in $C^L \setminus C^{>L}$ belong to the interior of W. Set $\bar{f} = \hat{f}|W$. We have by construction:

    a) $\bar{f}|W^L = f|W^L$;

    b) $0 \notin \bar{f}(\partial W)$;

    c) $0 \notin f(\partial W)$;

    d) for any disc $D(L, H) \subset W$ orthogonal to C^L and centered at a zero of $\bar{f}|W^L$ one has $\bar{f}(\partial D(L, H)) \cap B^L \neq \emptyset$.

From b) and c) we have that $\deg_0 \bar{f}$ and $\deg_0 f|W$ are correctly defined. As usual, we can obtain from our maps $f|W$ and $\bar{f}$ a pair of equivariant maps

$$f|\partial W, \bar{f}|\partial W : \partial W \to S_0$$

where S_0 is a small invariant sphere in B.

It follows from d) and Lemma 3.5(b) that $\deg_0 \bar{f} = 0$. Observe also that without loss of generality we can assume that ∂W is connected.

We are now in a position to treat our spurious orbit type (L). We have by constrution:

$$\deg \mu(L, H) \cdot b(L) \cdot |N(H, L)| = \deg_0 f|V(L) = \deg_0 f|W = \deg f|\partial W - \deg \bar{f}|\partial W.$$

Using (vi) and the assumption that there are not spurious orbit types between (H) and (L) we can apply the theorem we are proving to the pair $f|\partial W$ and $\bar{f}|\partial W$. We have:

$$\deg f|\partial W - \deg \bar{f}|\partial W = \deg \mu(L, H) \cdot b'(L) \cdot |N(H, L)| + \beta$$

where

$$\beta = \sum_{(K) \in O_{**}(M, S, H),\ (K) < (L)} \alpha(K, H) b'(K) |N(H, K)|.$$

And since

$$0 = \deg f|(\partial W)^L - \deg \bar{f}|(\partial W)^L = b'(L) \cdot |W(L)|$$

we must have that $b'(L) = 0$. Therefore,

$$\deg \mu(L, H) \cdot b(L)|N(H, L)| = \sum_{(K) \in O_{**}(M, S, H),\ (K) > (L)} \alpha(K, H) b'(K) |N(H, K)|,$$

and we are done with the induction step.

Summing up all the three cases we see that Theorem 3.1. follows from

Lemma 3.8. *There exists an equivariant map $F : C \to B$ which satisfies conditions (i) - (vi).*

3.4.3. To prove Lemma 3.8 we need one additional construction.

Take some $(H) \in O_*(M, S)$.

For each $B^K \subset B^H$, $(K) \in \mathrm{Or}(M)$, such that $\dim B^K < \dim C^K$ pick up a linear subspace $\tilde{B}_H^K$ satisfying the following conditions: (a) $B^K \subset B \cap \tilde{B}_H^K = B_H^K \subset B^H$; (b) $\dim B_H^K = \dim B^K + 1$, and (c) B_H^K does not coincide with any of the sets B^L, $(L) \in \mathrm{Or}(M)$, $(H) < (L)$.

Now with any $(H) \in O_*(M, S)$ we associate a family $\Lambda(H)$ which consists of all B^K and B_H^K $((K) \in \mathrm{Or}(M)$, $K \supset H)$.

Let U_H be a neighborhood of $M^{>H}$ in M^H which satisfies all the conditions of Lemma A (see also Remarks 3.2 and 3.3). Let $f_H : \partial U_H \to B^H$ be an equivariant map which satisfies the conditions of Lemma B with $G = W(H)$, $X = \partial U_H$, $B^{d+1} = B^H$, $k = 1$ and the family $\{L_j \cap B^{d+1}\}$, $j = 1, \ldots, m$, coinciding with $\Lambda(H)$.

Now we are in a position to present a basic construction. Take a subgroup K, $(K) \in \mathrm{Or}(M)$, $\dim B^K \leq \dim C^K$.

Set

$$Y(K) = \bigcup_{\{(H) \in O_*(M,S),\ (H)<(K)\}} \bigcup_{\{P \in \Lambda(H),\ \dim P \leq \dim B^K\}} \pi_H(f_H^{-1}(P)) \cap M^K$$

if $\dim C^K = \dim B^K$, and set

$$Y(K) = \bigcup_{\{(H) \in O_*(M,S),\ (H)<(K)\}} \bigcup_{\{P \in \Lambda(H),\ \dim P \leq \dim B^K+1\}} \pi_H(f_H^{-1}(P)) \cap M^K$$

if $\dim C^K > \dim B^K$.

In what follows we essentially use the following property of the sets $Y(K)$. Taking into account Remarks 3.2 and 3.3 and Lemma B we have

$$\dim(Y(K) \times I) = \dim Y(K) + 1 \leq \dim B^K - 1, \tag{3.6}$$

where I is the segment $[0, 1]$.

3.4.4. **Proof** of Lemma 3.8. We will construct a required map by induction over the orbit types as a series of extensions. At every step we begin extending a map $F : C^{>H} \to B^{>H}$ onto C^H by first considering isolated points of the set $F^{-1}(0)$. Let z be one of these points $(G_z = K)$. Take a sufficiently small closed trivial $W(K, H)$-invariant neighborhood $U_{z,C^H} = U_{z,C^K} \times D^l \subset C^H$ of z, where D^l is a disk orthogonal to C^K and centered at z. Take the unit disc $T(K, H)$ orthogonal to B^K in B^H.

Suppose that there exists a $W(K, H)$-equivariant extension

$$\varphi : \partial D^l \to \partial T(K, H)$$

of the map $F|(C^{>H} \bigcap \partial D^l) : C^{>H} \bigcap \partial D^l \to B^{>H} \bigcap \partial T(K, H)$ (for example, this is always the case when $(K) \in O_{**}(M, S, H)$, and then we use the map $f_{K,H}$ defined in Subsection 3.1.1 to construct φ). Further, using φ extend the map F $W(K, H)$-equivariantly over U_{z,C^H} in a natural way. Since the group $W(H)$ is finite, this leads to a $W(H)$-equivariant extension over $W(H)(U_{z,C^H})$ if the neighborhood U_{z,C^H} is sufficiently small.

Continuing in this fashion we get a $W(H)$-equivariant extension of the map F over a closed invariant set W which is obtained by adding to $C^{>H}$ small pieces of tubular neighborhoods of C^K. Clearly, this can be done in such a way that W is separated from the rest of $F^{-1}(0)$. Note, that the conditions (i)-(vi) are satisfied by the extension of F into W. We will denote the resulting extension by the same letter F.

Of course, an extension over the orthogonal sphere may not exist. Then we "forget" about the corresponding isolated "zero" at this stage.

The subsequent extension of F will depend on the ratio between $\dim C^H$ and $\dim B^H$. We consider several cases.

Case I. $\dim C^H > \dim B^H$.

Let $\tilde{Z} = F^{-1}(0) \setminus W \subset C^{>H}$ and $Q = C^{>H} \bigcup W(H)(Y(H) \times I)$. We want to extend F over C^H in such a way that the extension will not have zeros in $Q \setminus C^{>H}$. Take a decreasing chain of balls $B_i \subset B^H$ centered at the origin, and a decreasing sequence of closed invariant neighborhoods V_i of $\tilde{Z}$ in C^H such that

$$\bigcap_i B_i = 0, \qquad \bigcap_i V_i = \tilde{Z}, \qquad F(V_i \bigcap C^{>H}) \subset B_i, \qquad i = 1, 2, \dots$$

Let $Q_i = Q \setminus \text{Int}(V_i)$. It follows from (3.6) and Corollary 1.4 that the map F equivariantly extends to a map $\bar{F}_1 : C^{>H} \bigcup (\partial V_1 \bigcap Q) \to B^H$ in such a way that $\bar{F}_1(\partial V_1 \bigcap Q) \subset B_1 \setminus \{0\}$. By the same argument $\bar{F}_1$ can be further extended to a map $\tilde{F}_1 : C^{>H} \bigcup Q_1 \to B^H$ in such a way that $\tilde{F}_1(Q_1) \subset B^H \setminus \{0\}$. Using Corollary 1.4 once again extend the map $\tilde{F}_1$ to a map $F_1 : C^{>H} \bigcup Q \to B^H$ requiring this time that $F_1(V_1) \subset B_1$.

Suppose by induction that $F_i : C^{>H} \bigcup Q \to B^H$ is an equivariant extension of F, such that $F_i(V_i) \subset B_i$ and that $F_i(Q_i) \subset B^H \setminus \{0\}$. Continuing in this manner, take an equivariant extension of F_i over $V_{i+1} \bigcap Q$ which maps $\partial V_{i+1} \bigcap Q$ into $B_{i+1} \setminus \{0\}$. This extension together with the restriction of F_i onto Q_i form an equivariant map which by Corollary 1.4 yields a map $F_{i+1} : C^{>H} \bigcup Q \to B^H$ such that $F_{i+1}(V_{i+1}) \subset B_{i+1}$ and that $F_{i+1}(Q_{i+1}) \subset B^H \setminus \{0\}$. All this means that we have a uniformly bounded sequence $\{F_i\}$, $i = 1, 2, ...$, of $W(H)$-equivariant extensions of F which converges (uniformly) to an equivariant map $F_0 : C^{>H} \bigcup Q \to B^H$ such that $F_0(Q \setminus C^{>H}) \neq 0$.

To complete the inductive step in Case I extend F_0 over C^H equivariantly in an arbitrary way. Without loss of generality (see Lemma 1.18) one can assume that the set of zeros of the extension consists of finitely many connected components.

Note that the arguments presented above resemble those used in the proof of the well-known Urysohn Lemma.

Case II. $\dim C^H = \dim B^H$.

We extend F in three steps.

Step 1. Let Z, L, K_1, ..., K_s and Γ be as in (ii). We assume that Z is not an isolated point (see above). The goal of this step is to extend F over a neighborhood $U_Z \supset Z$ in such a way that the extension will satisfy condition (ii) from Lemma 3.8.

Let $U_H \subset M^H$ be an invariant neighborhood of $M^{>H}$ constructed in Lemma A. Take $\Lambda(H)$ defined in Subsection 3.4.3. Let $f_H : \partial U_H \to B^H$ be a map constructed in Lemma B ($\{L_j \bigcap B^{d+1}, \ j = 1, ..., m\} = \Lambda(H)$). Take $0 < a < b < 1$. Using the parametric family of curves $\gamma(x, t)$, $0 \le t \le 1$, $x \in \partial U_H$, described in Lemma A define the map F_0 which maps $P = U_H \times [a, b] \supset M^{>H} \times [a, b]$ into B^H by the rule:

$$F_0(< \gamma(x, t), s >) = F(< \gamma(x, 0), s >)(1 - t) + f_H(x)t.$$

Let F_1 be a restriction of F_0 to a sufficiently small closed neighborhood V of Z such that other connected components of "zeros" of the map F_0 do not belong to V. Let us check that F_1 satisfies condition (ii) .

Set

$$E = \bigcup_{\dim C^{K_i} > \dim B^{K_i}} C^{K_i}.$$

Take a point $p = < \gamma(x, t), s > \in P$ such that $(\bar{\pi}_H(x), s) = z \in Z$ (see Remarks 3.2 and 3.3). Suppose that $z \in Z \setminus E$ and let $F_1(p) \in B^L$ (respectively, $F_1(p) \in B_H^L$ if $\dim C^L > \dim B^L$ (see Subsection 3.4.3 for the definition of B_H^L)). By the construction of the map F_0 this means that $f_H(x) \in B^L$ (respectively, $f_H(x) \in B_H^L$). Hence (see the definition of $Y(L)$), $z \in Y(L) \times [a, b]$ which contradicts the construction of the map F. Therefore, $F_1(p) \notin B^L$ (respectively, $F_1(p) \notin B_H^L$) for any $p \in P$ such that $(\bar{\pi}_H(x), s) \in Z \setminus E$. Hence, there exists a neighborhood U_0 of Z in $C^{>H}$ such that $U_Z = \bar{\pi}_H^{-1}(U_0) \subset V$ and $(F_1|U_Z)^{-1}(B^L) \subset C^L \bigcup E$.

It is easy to see that $F_1|U_Z$ satisfies the remaining conditions of Lemma 3.5 as well.

Finally, by equivariance, we have an extension F_2 of our map over $\hat{U}_Z = W(H)(U_Z)$.

In a similar way one can treat the remaining "bad" connected components Z of Γ. Set $V_0 = \bigcup_Z \hat{U}_Z$, where the union is taken over "bad" connected components, and denote by F_3 the corresponding extension over $C^{>H} \bigcup W \bigcup V_0$ (here W is associated with isolated "zeros" (see above)).

Step 2. The next step is to extend the map F_3 over

$$C^{>H} \bigcup V_0 \bigcup W \bigcup (Y(H) \times [0,1])$$

using the construction described in Case I above.

Step 3. Finally, we use Corollary 1.5 to get an equivariant extension of our map over C^H.

Case III. $\dim C^H < \dim B^H$.

We proceed as in Case II using Corollary 1.4 instead of Corollary 1.5.

This completes the proof of Theorem 3.1.

We will complete this subsection with some additional comments.

Denote by $\overline{O}(M, N)$ the set of all orbit types $(K) \in O_*(M, N)$ such that there exists a K-equivariant map $\mu(K) : \sigma(K) \to \tau(K)$ where $\sigma(K)$ (respectively, $\tau(K)$) is a the unit sphere in the normal slice to M^K (respectively, to S^K) in M (respectively, in S). From the proof of Theorem 3.1 it follows immediately that for for any pair of G-equivarint maps $\Phi, \Psi : M \to S$ there exist integers $b(K)$, $(K) \in \overline{O}(M, N)$, such that

$$\deg \Phi - \deg \Psi = \sum_{(K) \in \overline{O}(M,S)} b(K) \cdot \deg \mu(K) \cdot |G/K|.$$

Clearly, $O_{**}(M, S) \subset \overline{O}(M, S)$. However, we don't know if $Q_{**}(M, S) = \overline{O}(M, S)$ (cf. Corollary 1.6). The analysis of the proof of Theorem 3.1 shows that if $(H) \in \overline{O}(M, S) \setminus O_{**}(M, S)$ then

$$b(H) \cdot \deg \mu(H) \cdot |G/H| \equiv 0 \pmod{d}$$

where $d = \text{GCD} \{|G/K|, (K) < (H), (K) \in O_{**}(M, S)\}$.

Let $d_1 = \text{GCD}\{|G/K|,\ (K) \in O_{**}(M,S)\}$, $d_2 = \text{GCD}\{|G/K|,\ (K) \in \overline{O}(M,S)\}$ and $d_3 = \text{GCD}\{|G/K|,\ (K) \in O_*(M,S)\}$.

Clearly (see above), d_2 devides d_1 and we don't know if $d_2 = d_1$. Anyway, we have the following

Corollary 3.12.

a) *For any pair of G-equivariant maps* $\Phi, \Psi : M \to S$

$$\deg \Phi \equiv \deg \Psi \quad (\text{mod } d_1)$$

and hence

$$\deg \Phi \equiv \deg \Psi \quad (\text{mod } d_2).$$

b) *Suppose that for any* $H \in O_*(M,S)$ *we have* $\dim M^K \leq \dim S^K$ *for all* $(K) < (H)$. *Then* $O_*(M,S) = \overline{O}(M,S) = O_{**}(M,S)$ *and hence* $d_1 = d_2 = d_3$.

3.5. The case of a compact Lie group

3.5.1. Proof of Corollary 3.3.

(a) $(T) \notin O_{**}(M,S)$.

Take a prime p and the integer i_1 provided by Proposition 3.1(b). Applying Corollary 3.2 to a subgroup G_{i_1} we get $\deg \Phi \equiv \deg \Psi \pmod{p}$. Since the prime number p is arbitrary, $\deg \Phi = \deg \Psi$ in this case.

(b) $(T) \in O_{**}(M,S)$.

Recall that under our assumptions M^T is connected. In addition, M^T is oriented (see [Bre]).

Take a prime p and the integer i_1 provided by Proposition 3.1(b). We can apply the arguments from the proof of Lemma 3.8 to the action of G_{i_1} with one modification. Namely, we define the number $\alpha(M,S,G_{i_1})$ as $\alpha(M,S,T)$ by constructing a T-equivariant map on the normal slice to $M^{G_{i_1}} = M^T$. Applying Corollary 3.2 to the G_{i_1}-actions we have

$$\deg \Phi - \deg \Psi \equiv (\deg(\Phi|M^T) - \deg(\Psi|M^T))\alpha(M,S,T) \quad (\text{mod } p).$$

Now taking into account that the prime p is arbitrary and that the definition of $\alpha(M,S,T)$ is independent of p, the proof of Corollary 3.3 is complete.

3.5.2. Proof of Theorem 3.2.

(a) If $(T) \notin O_{**}(M, S)$ then it follows from Corollary 3.3 that $\deg \Phi = \deg \Psi$.

(b) Using Corollary 3.3 once again we get

$$\deg \Phi - \deg \Psi = (\deg(\Phi | M^T) - \deg(\Psi | M^T))\alpha(T).$$

Further, we apply Corollary 3.4 (see also Remark 3.1) to M^T, getting

$$\deg(\Phi | M^T) - \deg(\Psi | M^T) = \sum_{(K) \in O_{**}(M, S, T)} \alpha(K, T)b(K)|N(T, K)|$$

and the result follows from

Lemma 3.9 (see [Bo2]). *If K is a subgroup of maximal rank of a compact Lie group G then $\chi(G/K) = |N(T, K)|$.*

3.6. Some special cases

3.6.1. Proof of Corollary 3.6. Let R^1 be a trivial one-dimensional representation of G and let $N_1 = R^1 \times N$. For $p \in N_1^G = R^1 \times N^G$ a tangent space $T_p(N_1)$ is a linear representation of G. Let S be a unit sphere in $T_p(N_1)$. It is clear that S is G-invariant and that $\dim S^H = \dim N^H$ for any $(H) \in \mathrm{Or}(N, G) = \mathrm{Or}(S, G)$. Therefore, by Theorem 1.3 there exists an equivariant map $\lambda : N \to S$. Suppose that we can pick up λ in such a way that $\deg \lambda = 1$. Then to prove the corollary it suffices to apply the corresponding "spherical" results to the map $\lambda \circ \Phi$, bearing in mind the definition of numbers $\alpha(K, H, N, S)$.

Thus we have proved the corollary up to the assumption that we can take $\lambda : N \to S$ so that $\deg \lambda = 1$. The existence of such λ follows from the Equivariant Hopf Theorem which will be proved in Chapter 5 (see Corollary 5.3(c)).

3.6.2.   Proof of Corollary 3.8. It is easy to see that under the conditions of the corollary the effective action of $W(K, H)$ on a sphere $\sigma_{L,H}$ normal to M^L in M^H (at some point $p \in M^L \subset M^H$) is semi-free with the fixed point set being a sphere $\sigma_{L,K}$ normal to M^L in M^K. Therefore the statement follows from Corollary 3.7 and the definition of the numbers α.

3.6.3. Proof of Corollary 3.9. Since we are assumed that G is abelian the factor group $\Gamma = (N(H) \bigcap L)/H = L/H$ acts on the normal slice V at a point $p \in M^L \subset M^H$. By the same reason $W = V^K$ is a Γ-invariant subspace of V which is Γ-equivariantly isomorphic to an L/K-invariant subspace normal to M^L in M^K. Similarly, the orthogonal complement U to V^K in V is Γ-invariant and is Γ-equivariantly isomorphic to a K/H-invariant subspace normal to M^K in M^H. Denote by $S(V)$, $S(W)$ and $S(U)$ the unit spheres in V, W and U respectively, and define Γ-equivariant maps $f_{L,K}$ on $S(W)$ and $f_{K,H}$ on $S(U)$. Take an equivariant join of $f_{L,K}$ and $f_{K,H}$. To complete the proof it remains to use the definition of the numbers α and the formula for an equivariant join degree (see Remark 1.2).

3.7. Historical and bibliographical notes

3.7.1. For equivariant maps which commute with a linear representation of a compact Lie group, Theorem 3.2 was established by E. Dancer [Dan] (see also [Rub]).

For (two) linear representations of a torus ($M^T = \emptyset$), Corollary 3.3 was established by L. Nirenberg [Ni2] and W. Marzantowicz [Mar1]. The case of maps from an invariant submanifold M of an S^1-module V into $V \backslash \{0\}$, $M^{S^1} \neq \emptyset$, was considered in [FHR]; the same situation for torus and finite group actions (see our Theorem 3.1) was considered by Wei-Yue Ding [Wei]. The case of two arbitrary linear representations of an abelian group was considered by J. Ize and A. Vignoli [IV] (see also [GKW], [Mar3]). Note also that in [Mar3] a formula expressing the degree of a map equivariant with respect to two linear representations of a compact Lie group is given in terms of the Euler classes of the representations (see also [Rab1]).

3.7.2. If G is a cyclic group and the second action is also semi-free Corollary 3.7(a) follows from the results due to Y. Izrailevich and E. Muchamadiyev [Iz,IM] (see also [BF]). For actions of a cyclic group on cohomological spheres a result similar to our Theorem 3.1 was obtained by T. Schelokova [Sc4] (see also [BF] where some results concerning arbitrary finite group actions on cohomological spheres were obtained). A connection between the so-called "equivariance index" introduced in [Sc4] (see also [BF,BoI,BoISc,Iz]) and the corresponding Euler class has been studied in [Mar3]). Corollary 3.5(e) for a finite p-group was obtained by W. Lück [Lü]. Constraints satisfied by the Lefschetz number of a map which commutes with an action of a compact Lie group G on a G-ENR space were considered by Komiya in [Kom] (see also [Rab2,Wan]). More detailed bibliographical information can be found in [St].

All the results of this chapter have been stated and proved by A. Kushkuley who gratefully acknowledges a significant help of Z. Balanov. In particular, Z. Balanov has pointed out several problems in the original manuscript.

Chapter 4

A WINDING NUMBER OF
EQUIVARIANT VECTOR FIELDS
IN INFINITE DIMENSIONAL
BANACH SPACES

To carry over the finite-dimensional results on degrees of equivariant maps to the infinite-dimensional case one has to overcome some significant technical difficulties. The most important of them (if one deals with two actions) is to construct a sufficiently large finite-dimensional space which is invariant with respect to a pair of linear actions. Therefore, in the first section we develop a machinery allowing us to solve this problem in some cases.

In the second section we present some results on winding numbers (Leray-Schauder degrees) of completely continuous (compact) equivariant vector fields. All the notions concerning degree theory in the infinite dimensional case one can find in [KZ], chapter 2; [Ni1], chapter 2.

4.1. Some auxiliary lemmas

4.1.1. In this subsection we introduce some notions which will be essentially used in what follows. Let $G^{(1)}$ and $G^{(2)}$ be two actions of a compact Lie group G in a Banach space E. By the same token for every $g \in G$ the symbols $g^{(1)}$ and $g^{(2)}$ denote the corresponding elements from $G^{(1)}$ and $G^{(2)}$. Let $\Omega \subset E$ be an open bounded region, $G^{(1)}(\overline{\Omega}) = \overline{\Omega}$, $N \subset \partial\Omega$ be a closed invariant subset and

$$\Phi = I - A, \ \Psi = I - B : (\overline{\Omega}, \partial\Omega) \to (E, E\backslash\{0\})$$

be a pair of compact vector fields (see, e.g., [KZ], p. 73). Suppose Φ and Ψ are equivariant with respect to $G^{(1)}$ and $G^{(2)}$. We say that Φ and Ψ are equivariantly homotopic on N if one can define a compact equivariant deformation joining $\Phi|N$ and $\Psi|N$.

Take $\varepsilon > 0$. We say that Φ is ε-equivariant if $||\Phi g^{(1)} - g^{(2)}\Phi|| < \varepsilon$ for every $g \in G$. In the same way one can define an ε-equivariant homotopy.

We also need the following definition. Let K and L be subspaces of E and $\delta > 0$. The subspace K is said to be δ-close to L if for every $x \in K \cap S$ there exists $x' \in L$ such that $||x - x'|| < \delta$ (cf. [Ka], p. 197); here S is the unit sphere in E.

4.1.2. In this subsection we generalize Theorem 3.2 to the case of ε-equivariant maps.

Let M and S be as in Theorem 3.2. Let G be a compact Lie group acting smoothly on M and orthogonally on S. Let $N \subset M$ be a closed invariant subset and $(H_1), \ldots, (H_m)$ the set of all orbit types in $M \backslash N$. Suppose $\varepsilon < 1$ is given and α is the Haar measure on G normed by the condition

$$\int_G d\alpha = 1.$$

Lemma 4.1 (see [KZ], Lemma 8.1). *Let* $\Phi, \Psi : M \to S$ *be a pair of* ε-*equivariant maps which are* ε-*equivariantly homotopic on* N. *Then the conclusions of "relative variants" of Theorems 3.1 and 3.2 are true.*

Proof. For every ε-equivariant map $\mu : M \to S$ ($\varepsilon < 1$) define a map $\mu_1 : M \to R^{n+1} \backslash \{0\}$ by the formula:

$$\mu_1(x) = \int_G g^{-1} \mu(gx) d\alpha.$$

One can easily see that μ_1 is non-degenerate and homotopic (in the space $(R^{n+1} \backslash \{0\})^M$) to μ. It suffices to check that $||\mu(x) - \mu_1(x)|| < 1$ for each $x \in M$. Indeed,

$$||\mu x - \mu_1(x)|| = ||\mu(x) - \int_G g^{-1} \mu g(x) d\alpha||$$

$$= || \int_G (\mu(x) - g^{-1} \mu g(x)) d\alpha||$$

$$\leq \int_G ||\mu(x) - g^{-1} \mu g(x)|| d\alpha$$

$$= \int_G ||g^{-1} g \mu(x) - g^{-1} \mu g(x)|| d\alpha$$

$$= \int_G (||g^{-1}|| \cdot ||g\mu(x) - \mu g(x)||) d\alpha$$

$$= \int_G ||g\mu(x) - \mu g(x)|| d\alpha < \varepsilon \cdot 1 = \varepsilon < 1.$$

Set $\hat{\mu} = \mu_1/||\mu_1||$; then $\hat{\mu}$ is homotopic to μ. Bearing in mind the linearity of the action of G on R^{n+1} it is easy to see that $\hat{\mu}$ is an equivariant map from M to S.

Now in order to prove Lemma 4.1 it suffices to note that the given maps Φ and Ψ are homotopic to the equivariant maps $\hat{\Phi}$ and $\hat{\Psi}$ respectively; if χ is an ε-equivariant homotopy joining $\Phi|N$ and $\Psi|N$ then $\hat{\chi}$ is an equivariant homotopy joining $\hat{\Phi}|N$ and $\hat{\Psi}|N$. It remains to use the relative variants of Theorems 3.1 and 3.2.

4.1.3.   In this subsection we prove Lemma 4.2 which in spite of the technical simplicity is crucial.

Lemma 4.2. *Suppose that E is a Hilbert space, $E_1 \subset E$ is a finite-dimensional subspace, $G^{(1)}$ and $G^{(2)}$ are isometrical actions of a finite group G and the operator $g^{(1)} - g^{(2)}$ is completely continuous for every $g \in G$. Then for any $\delta > 0$ there exists a finite-dimensional subspace $Q \subset E$ such that:*

1) $E_1 \subset Q$;
2) $g^{(1)}(Q) = Q$ for all $g \in G$;
3) $g^{(2)}(Q)$ is δ-close to Q for all $g \in G$.

Proof. For every $g \in G$ let us denote by A^g a linear operator of finite rank such that

$$||A^g - (g^{(1)} - g^{(2)})|| < \delta$$

and suppose the image of A^g is contained in a finite-dimensional subspace $T^g \subset E$. Let Q be the linear span of the set

$$\bigcup_{g \in G} g^{(1)}(E_1 + T^g).$$

Then inclusions 1) and 2) are obvious. Let us check 3). For given $x \in g^{(1)}(Q)$, where $||x|| = 1$, set $y = (g^{(2)})^{-1}(x)$. Then

$$x = g^{(2)}(y) = -(g^{(1)} - g^{(2)})(y) + g^{(1)}(y) = -A^g(y) + y_1 + g^{(1)}(y),$$

where $||y_1|| < \delta$. Setting $x' = -A^g(y) + g_1(y)$ we have (by the construction) that $x' \in Q$ and $||x - x'|| < \delta$.

4.1.4.   The sequence of Lemmas 4.3 - 4.5 which are proved below allows us to obtain the following

Proposition 4.1. *Let M be a finite-dimensional Euclidean space, $V : M \to M$ be an orthogonal operator which has an order q as an element of the transformation group. Suppose, further, that $K \subset M$ is a subspace, $\delta > 0$ is given and $V(K)$ is δ-close to K. Then there exists an orthogonal operator $V_0 : K \to K$ such that*
   1) $V_0^q = I$, *where I is the identity operator;*
   2) $\|V - V_0\| < 2\delta(2q + 1)$.

Lemma 4.3. *Let A and B be n-dimensional subspaces of some Euclidean space and B is δ-close to A. Then there exists an orthogonal operator $f : B \to A$ such that*

$$\|f - I\| < 2\delta. \tag{4.1}$$

Proof. Set $E = A + B$ and note that if $B \cap A_E^{\perp} \neq 0$ then B is δ-close to A with $\delta = 1$ and hence one can take any orthogonal operator for f (here $A_E^{\perp}$ is an orthogonal complement to A in E).

Let us prove, to begin with, that there exists an orthogonal basis $\bar{b} = (b_1, \ldots, b_n)$ in B such that its orthogonal projection on A is an orthogonal basis in A. Consider the case $A \cap B = 0$.

Let $\pi_1 : B \to A$ and $\pi_2 : B \to A_E^{\perp}$ be orthogonal projection operators. By assumption π_1 and π_2 are bijective operators. Let $\pi_0 : A \to A_E^{\perp}$ be an arbitrary non-degenerate orthogonal operator (since $\dim A = \dim A_E^{\perp}$ this operator is bijective). Define a map $\varphi : A \to A$ as a composition $\varphi := \pi_0^{-1} \pi_2 \pi_1^{-1}$ and consider a polar decomposition $\varphi = \mu\sigma$, where σ is a symmetric operator and μ is an orthogonal one. If $\bar{a}' = (a_1', \ldots, a_n')$ is an orthogonal basis in A, in which σ is diagonal, then $(\sigma(a_1'), \ldots, \sigma(a_n'))$ is an orthogonal basis in A; hence $(\mu(\sigma(a_1')), \ldots, \mu(\sigma(a_n')))$ is also an orthogonal basis in A from which it follows that

$$(\pi_2 \pi_1^{-1}(a_1'), \ldots, \pi_2 \pi_1^{-1}(a_n')) = (\pi_0 \varphi(a_1'), \ldots, \pi_0 \varphi(a_n'))$$

is an orthogonal basis in $A_E^{\perp}$. Set $a_i'' = \pi_2 \pi_1^{-1}(a_i')$, $b_i' = a_i' + a_i'' = \pi_1^{-1}(a_i')$ for all $i = 1, \ldots, n$. Since $(a_1', \ldots, a_n')$ and $(a_1'', \ldots, a_n'')$ are the orthogonal systems in A and in $A_E^{\perp}$ respectively, the system $(b_1', \ldots, b_n')$ is also orthogonal. Setting $b_i = b_i' / \|b_i'\|$ for all $i = 1, \ldots, n$ one can obtain an orthogonal basis in B whose orthogonal projection on A is an orthogonal basis.

In the remaining case set $A \cap B = C$, $A_1 = C_A^{\perp}$, $B_1 = C_B^{\perp}$. Then E can be decomposed as follows: $E = A_1 \oplus B_1 \oplus C$. Setting $E_1 = A_1 \oplus B_1$ and applying the above arguments to A_1 and B_1 one can obtain an orthogonal basis $(\tilde{b}_1, \ldots, \tilde{b}_m)$ in B_1 whose orthogonal projection on A_1 is an orthogonal basis. Since the orthogonal projection $B_1 \to A_1$ is a restriction of the orthogonal projection $B \to A$ one can obtain the required basis $(b_1, \ldots, b_n)$ complementing $(\tilde{b}_1, \ldots, \tilde{b}_m)$ with an arbitrary orthonormal basis in the subspace C.

Now if $(a'_1, \ldots, a'_n)$ is an orthogonal basis in A which is obtained as an orthogonal projection of the basis $(b_1, \ldots, b_n)$ whose existence is established above, then one can obtain the required isometry setting

$$f(b_i) = a'_i / \|a'_i\| = a_i \quad (i = 1, \ldots, n).$$

In order to check estimate (4.1) we prove the following fact. Let $\{b_i\}$ and $\{a_i\}$ be as above $(i = 1, \ldots, n)$. Then for every set of real numbers $\alpha_1, \ldots, \alpha_n$ the system $(b_1 + \alpha_1 a_1, \ldots, b_n + \alpha_n a_n)$ is orthogonal.

Indeed, suppose $A \cap B = \emptyset$. Every b_i can be decomposed into the sum: $b_i = \tilde{a}_i + c_i$, where $c_i \in A_E^\perp$, $b_i = b'_i / \|b'_i\|$, $\tilde{a}'_i = a'_i / \|b'_i\|$, $c_i = a''_i / \|b'_i\|$, from which it follows that $(c_1, \ldots, c_n)$ is an orthogonal system. Further, the system $((1 + \alpha_1)a_1, \ldots, (1 + \alpha_n)a_n)$ is orthogonal and hence the system $(b_1 + \alpha_1 a_1, \ldots, b_n + \alpha_n a_n)$ is also orthogonal because $b_i + \alpha_i a_i = (1 + \alpha)a_i + c_i$. The case $A \cap B = C$ can be considered in a similar way.

We are now in a position to show (4.1). Let $w \in B$, $\|x\| = 1$. Expand x over the basis $(b_1, \ldots, b_n)$:

$$x = \beta_1 b_1 + \ldots + \beta_n b_n, \quad \sum_{i=1}^{n} \beta_i^2 = 1.$$

Then $f(x) = \beta_1 a_1 + \ldots + \beta_n a_n$, from which (in accordance with the orthogonality of system $(b_1 - a_1, \ldots, b_n - a_n)$) it follows:

$$\|f(x) - x\| = \left(\sum_{i=1}^{n} \beta_i^2 \|b_i - a_i\|^2 \right)^{1/2}$$

$$< \left(\sum_{i=1}^{n} \beta_i^2 (2(1 - (1 - \delta)^{1/2})) \right)^{1/2}$$

$$< 2\delta \left(\sum_{i=1}^{n} \beta_i^2 \right)^{1/2}$$

$$= 2\delta \cdot 1 = 2\delta.$$

The lemma is proved.

4.1.5.   Lemma 4.4. *Suppose that under the assumptions of Lemma 4.1* $f :$ $V(K) \to K$ *is an orthogonal operator provided by Lemma 4.3. Then for every* $x \in K$ *the following estimate is true:*

$$||(fV)^q x - x|| < 2q\delta ||x||.$$

Proof. Using an induction over m let us show that for every $x \in K$

$$||(fV)^m x - V^m x|| < 2m\delta ||x||.$$

For $m = 1$ we have:

$$||fVx - Vx|| = ||(f - I)Vx|| \leq ||f - I|| \cdot ||Vx|| < 2\delta ||x||.$$

Suppose, further, that for some $m \in N$ the following estimate is true:

$$||(fV)^m x - V^m x|| < 2m\delta ||x|| \quad (x \in K).$$

Then

$$
\begin{aligned}
||(fV)^{m+1} x - V^{m+1} x|| &= ||(fV)^{m+1} x - fV^{m+1} x + fV^{m+1} x - V^{m+1} x|| \\
&\leq ||(fV)^{m+1} x - fV^{m+1} x|| + ||fV^{m+1} x - V^{m+1} x|| \\
&= ||(fV)(fV)^m x - (fV)V^m x|| + ||(f - I)V^{m+1} x|| \\
&\leq ||fV|| \cdot ||(fV)^m x - V^m x|| + ||f - I|| \cdot ||V^{m+1} x|| \\
&= ||(fV)^m x - V^m x|| + ||f - I|| \cdot ||x|| \\
&< 2m\delta ||x|| + 2\delta ||x|| = 2(m + 1)\delta ||x||.
\end{aligned}
$$

Now setting $m = q$ and bearing in mind that $V^q x = x$ for every $x \in K$ one obtains the required estimate.

4.1.6.   Lemma 4.5. *Let K be a finite-dimensional Euclidean space, $V : K \to K$ be an orthogonal operator such that* $||V^p - I|| < \delta$ ($\delta > 0, p \geq 2$). *Then there exists an orthogonal operator $V_0 : K \to K$ such that:*
  1) $V_0^p = I$;
  2) $||V - V_0|| \leq 2\delta$
(if $\delta < 2$ then the inequality is strict).

Proof. Consider first the case $\delta < 2$. Let $V = V_1 \oplus V_2 \oplus \ldots \oplus V_k$ be a canonical decomposition of the operator V (see, for instance, [Ge], p. 124) and let $V_i = A_\varphi$

be a two-dimensional operator of rotation by φ. Define $\varphi' = 2\pi/p \cdot [p\varphi/2\pi]$, where $[a]$ means the nearest integer to a, and set $V_i^{(0)} = A_{\varphi'}$. Then

$$\|A_\varphi x - A_{\varphi'} x\| \leq 2 \cdot |\sin((\varphi - \varphi')/2)| \cdot \|x\|$$
$$\leq 2 \cdot |\varphi - \varphi'|/2 \cdot \|x\|$$
$$= |\varphi - \varphi'| \cdot \|x\|,$$

i.e.

$$\|V_i - V_i^{(0)}\| \leq |\varphi - \varphi'|.$$

Since by assumption $\|V^p - I\| < \delta$ then $\|V_i^p - I\| < \delta$. On the other hand,

$$\|V_i^p - I\| = \|A_\varphi^p - I\| = \|A_{p\varphi} - I\| = 2|\sin((p\varphi)/2)|.$$

Hence (using the inequality $|\sin(\alpha)/\alpha| \geq 2/\pi$ $(0 < \alpha < \pi/2)$) we have:

$$\delta > 2 \cdot |\sin(p\varphi/2)| = 2 \cdot |\sin(p/2(\varphi - \varphi' + \varphi'))|$$
$$= 2 \cdot |\sin(p/2(\varphi - \varphi') + p/2 \cdot 2\pi/p \cdot [p\varphi/2\pi])|$$
$$= 2 \cdot |\sin(p/2(\varphi - \varphi'))|$$
$$\geq 2 \cdot p/2|\varphi - \varphi'| \cdot 2/\pi$$
$$= 2p/\pi|\varphi - \varphi'|,$$

from which it follows that

$$|\varphi - \varphi'| < \pi\delta/2p \quad \text{and} \quad \|V_i - V_i^{(0)}\| < \pi\delta/2p.$$

Let operator V_0 be obtained from operator V by replacing V_i (in its canonical decomposition) with $V_i^{(0)}$ acting on the same subspace; the one-dimensional blocks of the matrix of the operator V are not changed by this procedure.

Then in accordance with the above arguments $\|V - V_0\| < 2\delta$. By the construction $(V_i^{(0)})^p = I$. It is clear that $V_0^p = I$ if the matrix of the canonical decomposition of V does not contain (1×1)-blocks. If it has such blocks then by the inequality $\|V^p - I\| < 2$ we can conclude that p is even and then $V_0^p = I$ once again.

In the case when $\delta \geq 2$ it suffices to set $V_0 = I$. Then $\|V - V_0\| \leq \|V\| + \|V_0\| = 2$.

The lemma is completely proved.

4.1.7. Proof of Proposition 4.1.

a) By Lemma 4.3 there exists an orthogonal operator $f : V(K) \to K$ such that $\|f(x) - x\| < 2\delta\|x\|$. According to Lemma 4.4 $\|fV - V\| < 2\delta$ and $\|(fV)^q - I\| < 2q\delta$. Finally, by Lemma 4.5 there exists an orthogonal operator $V_0 : K \to K$ such that $V_0^q = I$ and $\|fV - V_0\| < 4q\delta$ from which it follows that

$$\|V - V_0\| = \|V - fV + fV - V_0\| \leq \|V - fV\| + \|fV - V_0\| < 2\delta(2q + 1).$$

4.2. Main results

4.2.1.   In this section we study the winding number of equivariant compact vector fields in infinite-dimensional Banach spaces.

Let us introduce the following

Definition.   *Suppose* Ω *is a bounded open region in a Banach space* E. *The boundary of* Ω *is said to be an* M-*manifold if for any sufficiently large finite-dimensional subspace* $E_1 \subset E$ *the intersection* $E_1 \cap \partial\Omega$ *is a connected oriented smooth submanifold.*

A sphere in a Banach space is a typical example of an M-manifold.

In this subsection we shall prove

Theorem 4.1. *Let* E *be a Hilbert space where two isometrical linear actions* $G^{(1)}$ *and* $G^{(2)}$ *of a finite cyclic group* G *are given* $(|G| = p)$. *Suppose* Ω *is a bounded open region in* E, $G^{(1)}(\overline{\Omega}) = \overline{\Omega}$ *and* $\partial\Omega$ *is an* M-*manifold. Finally, suppose* $N \subset \overline{\Omega}$ *is a closed* $G^{(1)}$-*invariant subset and* G *acts on* $\partial\Omega \backslash N$ *with one orbit type* (H).

Then for any pair of compact vector fields

$$\Phi = I - A, \ \Psi = I - B : (\overline{\Omega}, \partial\Omega) \to (E, E\backslash\{0\})$$

which are G-*equivariantly homotopic on* $(N, \partial\Omega \cap N)$ *the following formula is true:*

$$\gamma(\Phi) \equiv \gamma(\Psi) \pmod{|G/H|}$$

(here $\gamma(\Phi)$ *and* $\gamma(\Psi)$ *are the winding numbers of the vector fields* Φ *and* Ψ *on* $\partial\Omega$).

Proof. Let $x - \mu(x,t)$, $(x \in N, \ t \in [0,1])$ be a compact equivariant deformation non-degenerated on $N \cap \partial\Omega$ (its existence is provided by the theorem assumptions). Bearing in mind the compactness of the vector fields Φ and Ψ one can state that these fields as well as the homotopy $I - \mu$ are uniformly non-degenerate on $\partial\Omega$ and $N \cap \partial\Omega$ respectively (see [KZ], Lemma 19.1).

For any fixed $\varepsilon_1 > 0$ there exist finite-dimensional vector fields $\Phi_1, \Psi_1 : (\overline{\Omega}, \partial\Omega) \to (E, E\backslash\{0\})$ and a finite-dimensional deformation $I - \mu_1 : (N, \partial\Omega \cap N) \times [0,1] \to (E, E\backslash\{0\})$ such that

$$||\Phi - \Phi_1|| < \varepsilon_1; \quad ||\Psi - \Psi_1|| < \varepsilon_1; \quad ||\mu - \mu_1|| < \varepsilon_1.$$

Suppose the images of the operators $I - \Phi_1$, $I - \Psi_1$, $I - \mu_1$ belong to a finite-dimensional subspace $E_1 \subset E$. Without loss of generality one can assume that E_1

contains at least one interior point of Ω. Then taking into account the equivariance of Φ and Ψ on $\overline{\Omega}$ one can conclude (Lemma 4.2) that for any $\delta > 0$ there exists a finite-dimensional subspace $Q \subset E$ such that:

1) $E_1 \subset Q$;

2) $U(Q) \subset Q$;

3) $V(Q)$ δ-close to V

(here U and V are generators of $G^{(1)}$ and $G^{(2)}$ respectively). By Proposition 4.1 there exists an orthogonal operator $V_0 : Q \to Q$ such that:

1) $V_0^p = I$;

2) $\|V - V_0\| < 2\delta(2q + 1)$;

Thus due to the suitable choice of δ the operator V_0 is close to V as much as desired. Let $\pi : E \to Q$ be the orthogonal projection operator. For every $x \in \partial\Omega$ set $\Phi_2 x = x - \pi A x$; $\Psi_2 x = x - \pi B x$; for every $x \in \partial\Omega \cap N$ and $t \in [0,1]$ set $\mu_2(x,t) = \pi\mu(x,t)$. Note also that

$$\|\Phi - \Phi_2\| < \varepsilon_1, \quad \|\Psi - \Psi_2\| < \varepsilon_1, \quad \|\mu - \mu_2\| < \varepsilon_1.$$

It is clear that $I - \mu_2$ is a homotopy between $\Phi_2|(N \cap \partial\Omega)$ and $\Psi_2|(N \cap \partial\Omega)$. It is obvious that due to the suitable choice of δ and ε_1 one can obtain ε_2-equivariance of Φ_2, Ψ_2 and $I - \mu_2$ for any ε_2. Let us restrict the fields Φ_2 and Ψ_2 on $Q \cap \partial\Omega$, and the homotopy $I - \mu_2$ on $N \cap \partial\Omega \cap Q$. Set $\Phi_3(x) = \Phi_2^*(x)/\|\Phi_2^*(x)\|$; $\Psi_3(x) = \Psi_2^*(x)/\|\Psi_2^*(x)\|$; $x - \mu_3(x,t) = (x - \mu_2^*(x,t))/\|(x - \mu_2^*(x,t)\|$, where Φ_2^*, Ψ_2^* and μ_2^* are the relevant restrictions on $Q \cap \partial\Omega$ and $N \cap Q \cap \partial\Omega$ respectively. Since $\Phi|\partial\Omega$, $\Psi|\partial\Omega$ and $(I - \mu)|(N \cap \partial\Omega)$ are uniformly non-degenerate, due to the suitable choice of ε_2 one can obtain an ε_3-equivariance of Φ_3, Ψ_3 and $I - \mu_3$, where ε_3 is an arbitrary positive number. Setting now $\varepsilon_3 = 1/2$ we obtain that $\Phi_3, \Psi_3 : \partial\Omega \cap Q \to S \cap Q$ are $1/2$-equivariant and, moreover, $\Phi_3|(N \cap Q \cap \partial\Omega)$ and $\Psi_3|(N \cap Q \cap \partial\Omega)$ are $1/2$-equivariantly homotopic, i.e. (Lemma 4.1)

$$\gamma(\Psi, \partial\Omega) = \gamma(\Psi_3, \partial\Omega \cap Q) \equiv \gamma(\Phi, \partial\Omega) = \gamma(\Phi_3, \partial\Omega \cap Q) \quad (\text{mod } |G/H|)$$

(here S is the unit sphere in E).

The theorem is proved.

Below we present some obvious consequences of Theorem 4.1 (cf. Chapter 2).

Corollary 4.1. *Suppose that under the assumptions of Theorem 4.1, $\partial\Omega\backslash N$ is a free G-space. Then*

$$\gamma(\Psi) \equiv \gamma(\Phi) \qquad (\text{mod } |G|).$$

Corollary 4.2. *Assume that under the assumptions of Theorem 4.1 G is an (infinite) group algebraically isomorphic to a compact non-completely disconnected group (cf. Lemma 2.3 and Corollary 2.9). Let $\partial\Omega \setminus N$ be a free G-space. Then*

$$\gamma(\Psi) = \gamma(\Phi).$$

4.2.2. Suppose now that E is a Banach space where an isometrical representation of a compact Lie group G is given. Let $\Omega \in E$ be an invariant open region, $0 \in \Omega$. For the sake of simplicity suppose that for any $(H) \in \text{Or}(\partial\Omega)$ (H is of maximal rank if G is infinite) the set $\partial\Omega \cap E^H$ is an M-manifold. Let Φ, $\Psi : (\overline{\Omega}, \partial\Omega) \to (E, E\backslash\{0\})$ be a pair of compact vector fields commuting with the action of G. Then using the scheme of the proof of Theorems 3.1 and 4.1 one can essentially strengthen Theorem 4.1. To give the precise formulation we need some notations.

If $\partial\Omega \cap E^G = \emptyset$ set $q = \text{GCD}\{\chi(G/K) \mid (K) \in \text{Or}(\partial\Omega)\}$ (K is assumed to be of maximal rank if G is infinite).

If $\partial\Omega \cap E^G \neq \emptyset$ set $q' = \text{GCD}\{\chi(G/K) \mid (K) \in \text{Or}(\partial\Omega), (K) \neq (G)\}$ (K is assumed to be of maximal rank if G is infinite).

If $\partial\Omega \cap E^G \neq \emptyset$ and $(H) \in \text{Or}(\partial\Omega)$ (and H is of maximal rank if G is infinite) set $q_H = \text{GCD}\{|N(H,K)| \mid (K) \geq (H), (K) \neq (G)\}$.

Theorem 4.2 (cf. Theorems 3.1 and 3.2). *Under the assumptions above*
    (a) *if $\partial\Omega \cap E^G = \emptyset$ then*

$$\gamma(\Phi) \equiv 1 \pmod{q};$$

in particular, if $q > 1$ then $\gamma(\Phi) \neq 0$.

    (b) *If G is infinite and $\text{Or}(\partial\Omega)$ does not contain any subgroup of maximal rank then*

$$\gamma(\Phi) = 1;$$

    (c) *if $\partial\Omega \cap E^G \neq \emptyset$ then*

$$\gamma(\Phi) \equiv \gamma(\Phi|(\partial\Omega \cap E^G)) \pmod{q'};$$

in particular, if $\gamma(\Phi|(\partial\Omega \cap E^G)) = \pm 1$ and $q' > 1$ then $\gamma(\Phi) \neq 0$;

*(d) if $(H) \in \mathrm{Or}(\partial\Omega)$ (with H of maximal rank if G is infinite) and $\partial\Omega \cap E^G \neq \emptyset$
then*

$$\gamma(\Phi|(\partial\Omega \cap E^H)) \equiv \gamma(\Phi|(\partial\Omega \cap E^G)) \pmod{q_H}.$$

Remark 4.1. It is easy to see that one can carry over the results of Chapter 3 to the infinite dimensional case considering the following situation. Let G be a compact Lie group acting isometrically on a Banach space E and finite-dimensional linear spaces V and W ($\dim V = \dim W$). Let $\bar\Omega \subset V \times E$ be as above and let $f : (\bar\Omega, \partial\Omega) \to (W \times E, W \times E \setminus \{0\})$ be an equivariant map of the form $f(x,y) = (f_1(x,y), y - f_2(x,y))$ where f_2 is compact (cf. [IV]). We omit the obvious reformulations.

4.3. Historical and bibliographical notes

4.3.1. Note that Theorem 4.1 is valid in arbitrary Banach spaces if the closure of the transformation group $< U, V >$ where U and V are the generators of $G^{(1)}$ and $G^{(2)}$ respectively, is compact. In particular, if U and V commute, then the above mentioned group is finite.

Under the assumptions that U and V commute, the first action is free on $\partial\Omega$, $N = \emptyset$ and $\partial\Omega$ is a sphere Theorem 4.1 was proved (in Banach space) by E. Muhamadiev [Mu].

In the case when A and B are smooth, both actions are free (on $\partial\Omega$ and $E\setminus\{0\}$ relatively) and $N = \emptyset$ the statement of Theorem 4.1. (in Banach space) follows from the formula for equivariant Fredholm operators obtained by V. Zvjagin, E. Muhamadiev and Yu. Sapronov [ZMS]; see also V. Zvjagin's paper [Zv] where the situation of a pair of smooth actions was considered. In the given formulation Theorem 4.1 was proved by Z. Balanov and S. Vinichenko [BV2] (see also [Ba2], [Ba3]). The case when A (and B) satisfy the condition $AU = VA$ and $\partial\Omega$ is a sphere, was considered by Y. Izrailevich and V. Obuhovsky in [IO1].

4.3.2. For the antipodal action on a sphere Theorem 4.2(a) was proved by M. Krasnoselsky [Kr2]. If G is a finite cyclic group acting freely on a sphere Theorem 4.2(a) was proved by E. Muhamadiev [Mu] (see also IO2]). If G is a finite group or a torus Theorem 4.2(a,b,c) was established by Wei-Yue Ding [We] (see also [Bar2]). For an arbitrary compact Lie group representation ($\partial\Omega$ is a unit sphere) Theorem 4.2(a,b,c) was proved by E.N. Dancer [Dan].

For abelian group actions the situation described in Remark 4.1 has been studied in detail in [IV] under the assumption that $\dim V^H = \dim W^H$ for all $H \in \mathrm{Iso}(\bar{\Omega})$ (actually, the authors considered a more complicated situation with $R^k \times V$ instead of V with the trivial G-action on R^k).

For a detailed bibliography on the degree of equivariant maps in the infinite dimensional case we refer to [St].

Chapter 5

SOME APPLICATIONS

In this chapter we present some applications of the results obtained in the previous chapters.

In the first section we consider an elliptic semi-linear boundary value problem which is associated with the corresponding linear problem of positive Fredholm index. Under some symmetry assumptions we prove the existence of solutions of arbitrarily large norm in the corresponding Hölder space. We follow the scheme by P. Rabinowitz [Ra1] (see also [Mar1]).

In the second section we give a lower estimate for the genus of a free part of a finite-dimensional sphere S with a compact Lie group action. To treat this problem we modify the well-known geometric aproach by M. Krasnoselskii [Kr1, KZ]. The obtained result is applied to the irreducible $SO(n)$-representations in spherical harmonics. Studying this situation is motivated by the bifurcation phenomena investigations in semilinear boundary-value problems on a ball (see, for instance, [Bar1]).

In the third section we present an equivariant version of the Hopf Theorem on the homotopy classification of mappings from a manifold to a sphere. In contrast to [Di1, Di2] (see also [To], [Lai]) we do not use any equivariant cohomology theory. Some illustrative examples also are considered.

The fourth section is devoted to the generalization of the Dold Theorem on the non-existence of an equivariant mapping from an n-dimensional free G-sphere to an m-dimensional one if $n > m$; we consider a situation of non-free actions on manifolds.

In the fifth section we give an elementary proof of the well-known theorem of M. Atiyah and D. Tall [AT] on p-group representations, and present some generalizations of this result related to Corollary 3.2.

5.1. Unbounded branches of solutions of elliptic equations

5.1.1.   Let us consider a linear elliptic boundary-value problem

$$Lu(x) \equiv \sum_{|\sigma| \leq 2m} a_\sigma(x) D^\sigma u(x) = 0, \quad x \in \Omega \,,$$

$$B_i u(x) \equiv \sum_{|\sigma| \leq m_i} b_{i\sigma}(x) D^\sigma u(x) = 0, \quad x \in \partial\Omega \,, \tag{5.1}$$

where $1 \leq i \leq m$, $m_i < 2m$, $\overline{\Omega}$ is the closure of the bounded region Ω in R^n with C^∞-smooth boundary, L is an elliptic operator with C^∞-smooth coefficients $a_\sigma : \Omega \to \mathrm{Hom}\ (R^q, R^q)$, where the Euclidean space R^q is the range of values of the function $u(\cdot)$. Let $\{B_i\}$ be a family of differential operators with C^∞-smooth coefficients $b_{i\sigma} : \partial\Omega \to \mathrm{Hom}\ (R^q, R^q)$ and let $|\sigma| = \sigma_1 + \ldots + \sigma_n$ be the norm of the multi-index $\sigma = (\sigma_1, \ldots, \sigma_n)$ (see [ADN], [Ni1], [Mar1], [Ra1]). Suppose our boundary conditions $B = \{B_i\}$ are "well posed" (see [Ni1], p. 43); in particular, this means that the pair (L, B) defines a Fredholm operator in the corresponding Hölder space. Denote the index of (L, B) by ν and suppose that $\nu > 0$. Then the set of solutions of problem (5.1) is a linear subspace whose dimension is positive. In particular, this means that for any $r \geq 0$ there exists a solution u_0 of problem (5.1) such that $\|u_0\| = r$.

Consider now a non-linear elliptic boundary-value problem which is associated with problem (5.1):

$$Lu(x) = \varphi(x, u(x), Du(x), \ldots, D^{2m-1} u(x)), \quad x \in \Omega \,,$$

$$B_i u(x) = 0, \quad x \in \partial\Omega \,, \tag{5.2}$$

where φ is a smooth function.

Our goal is to show that some symmetry conditions on a_σ, $b_{i\sigma}$ and φ provide the existence of solutions to problem (5.2) which have an arbitrarily large norm in the corresponding Hölder space.

5.1.2.   Here we present some known facts and describe the situation studied below.

If Ω is as above denote by $C^j(\overline{\Omega}, R^q)$ the space of j-smooth R^q-valued functions defined on $\overline{\Omega}$. For any $0 < \mu < 1$ let us denote by $C^{j+\mu}(\overline{\Omega}, R^q)$ the space of all $u \in C^j(\overline{\Omega}, R^q)$ such that

$$\|u\|_{j+\mu} = \|u\|_j + \sum_{|\sigma|=j} \sup_{\substack{x \neq y \\ x,y \in \Omega}} \frac{\|D^\sigma u(x) - D^\sigma u(y)\|}{\|x - y\|^\mu} \tag{5.3}$$

is finite, where

$$\|u\|_j = \sum_{|\sigma| \leq j} \max_{x \in \overline{\Omega}} \|D^\sigma u(x)\|$$

(see, for example [Ni1], p. 44).

Formula (5.3) defines the well-known Hölder norm in $C^{j+\mu}(\overline{\Omega}, R^q)$.

Suppose an orthogonal action of a compact Lie group G on R^q is given and assume the general common divisor of the Euler characteristics of the orbits for this action on the unit sphere in $R^q \ominus (R^q)^G$ is greater than one. Formula $(gu)(x) = g(u(x))$ for $g \in G$ defines the structure of a Banach G-space on $C^{j+\mu}(\overline{\Omega}, R^q)$.

We assume the matrices $a_\sigma(x)$ and $b_{i\sigma}(x')$ commute with the action of G on R^q for all $x \in \Omega$ and $x' \in \partial\Omega$, respectively. It is easy to see that this is a necessary and sufficient condition for operators L and B to be equivariant in the relevant functional spaces (see, for example, [Mar1]).

We suppose that the Fredholm index of the linear operator $P = (L, B)$, where

$$P : \{u \in C^{2m+\mu}(\overline{\Omega}, R^q); \quad Bu = 0 \text{ on } \partial\Omega\} \to C^\mu(\overline{\Omega}, R^q) \,,$$

is positive and, in addition that there exists a G-equivariant linear embedding of the co-kernel into the kernel (clearly, the last condition holds automatically if the co-kernel is trivial).

Lemma 5.1 (see [ADN], [Ra1], [Ni1], [Mar1]).
(i) Let $f \in C^\mu(\overline{\Omega}, R^q)$. If u is a solution of the problem

$$Lu(x) = f(x), \quad x \in \Omega \,,$$

$$Bu(x) = 0, \quad \in \partial\Omega$$

then $u \in C^{2m+\mu}(\overline{\Omega}, R^q)$.
(ii) For every function $u \in C^{2m+\mu}(\overline{\Omega}, R^q) \ominus \ker P$ the following estimate

$$\|u\|_{2m+\mu} \leq C \cdot \|Pu\|_\mu$$

holds, where C depends on P, μ and does not depend on u.

For the sake of simplicity we suppose that the restriction

$$P : \{u \in C^{2m+\mu}(\overline{\Omega}, R^q), \quad Bu = 0 \text{ on } \partial\Omega\}^G \to (C^\mu(\overline{\Omega}, R^q))^G$$

of the operator P is invertible.

Denote the space of k-linear operators from R^n to R^q by $L^k(R^n, R^q)$. Since G acts trivially on $\overline{\Omega}$, the diagonal action of G on

$$T = \overline{\Omega} \times R^q \times L^1(R^n, R^q) \times \ldots \times L^{2m-1}(R^n, R^q)$$

is correctly defined. Suppose $\varphi : T \to R^q$ is equivariant; then the formula

$$(\phi(u))(x) = \varphi(x, u(x), Du(x), \ldots, D^{2m-1}u(x))$$

defines an equivariant map

$$\phi : C^{2m-1+\mu}(\overline{\Omega}, R^q) \to C^{\mu}(\overline{\Omega}, R^q) .$$

Assume finally that there exist numbers $\delta > 0$ and $0 < \gamma < 1$ such that for every $u \in C^{2m-1+\mu}(\overline{\Omega}, R^q)^G$ the following inequality is true:

$$\|\varphi(x, u(x), Du(x), \ldots, D^{2m-1}u(x))\|_{R^q} < \delta(1 + \sum_{|\sigma| \leq 2m-1} \|D^\sigma u(x)\|_{R^q})^\gamma . \quad (5.4)$$

We prove the following

Theorem 5.1. *Under the above assumptions with respect to L, B, φ and G there exists $r_0 \geq 0$ such that for every $r \geq r_0$ problem (5.2) has a solution $u^{(0)} \in C^\infty(\overline{\Omega}, R^q)$ and $\|u^{(0)}\|_{2m-1+m} = r$. If G acts on R^q without non-zero G-fixed points then $r_0 = 0$.*

5.1.3.   Proof of Theorem 5.1. Denote the kernel span $\{v_1, \ldots, v_k\}$ of the operator P by V_k, and its co-kernel span $\{w_1, \ldots, w_d\}$ by W_d, where $v_1, \ldots, v_k$ and $w_1, \ldots, w_d$ are the corresponding linearly independent systems. Set $E = C^{2m-1+\mu}(\overline{\Omega}, R^q)$. Since V_k is isomorphic to R^k and W_d is isomorphic to R^d, we can write

$$E = R^k \times \overline{E} = R^d \times \tilde{E} , \quad (5.5)$$

where $\overline{E}$ and $\tilde{E}$ are orthogonal complements of R^k and R^d correspondingly with respect to the inner product of the space $L_2(\overline{\Omega}, R^q)$; this means that $h \in \overline{E}$ if and only if $< h, v_i >= 0$ with respect to $L_2(\overline{\Omega}, R^q)$, where $i = 1, \ldots, k$. Thus if $u \in E$ then

$$u = \sum_{i=1}^{k} \overline{a}_i v_i + \tilde{u} = \sum_{s=1}^{d} \tilde{a}_s w_s + \tilde{u}$$

or $u = (\overline{a}, \overline{u}) = (\tilde{a}, \tilde{u})$.

Let the operator $\tilde{P} : E \to \hat{E}$ be an orthogonal projection with respect to $L_2(\overline{\Omega}, R^q)$ defined by the formula $\tilde{P}u = \tilde{u}$. Consider an additional non-homogeneous problem

$$
\begin{aligned}
L\overline{U}(x) &= \tilde{P}f(x), \quad x \in \Omega , \\
B\overline{U}(x) &= 0, \quad x \in \partial\Omega ,
\end{aligned}
\tag{5.6}
$$

which is associated with problem (5.1), where $f \in E$. By Lemma 5.1(i) there exists a unique $\overline{U}_0 \in C^{2m+\mu}(\overline{\Omega}, R^q) \cap \overline{E}$ satisfying (5.6).

Rewrite (5.2) as an operator equation in E. Let $u = (\bar{a}, \bar{u}) \in E$ and $Fu = (\overline{A}, \overline{U}) = (A(u), Q(u))$, where $\overline{A} = (\overline{A}_1, \ldots, \overline{A}_k)$ is defined by the formula:

$$
\overline{A}_i = \begin{cases} \bar{a}_i - \langle \varphi(\cdot, u, Du, \ldots, D^{2m-1}u), w_i \rangle & 1 \le i \le d \\ \bar{a}_i & d+1 \le i \le k \end{cases}
\tag{5.7}
$$

(here $k > d$ since ind $(L, B) > 0$), operator Q is defined in accordance with (5.6) as the unique solution of the system

$$
\begin{aligned}
L\overline{U}(x) &= \tilde{P}\varphi(x, u(x), Du(x), \ldots, D^{2m-1}u(x)), \quad x \in \Omega \\
B\overline{U}(x) &= 0, \quad x \in \partial\Omega .
\end{aligned}
\tag{5.8}
$$

By (5.5), (5.6) and (5.8) the operator F maps the space E into itself. Since $a_\sigma(x)$, $b_{i\sigma}(x)$ and φ are equivariant and there exists an equivariant embedding of the co-kernel of P into its kernel, one can show that F is equivariant. In addition, combining statement (ii) of Lemma 5.1 with the smoothness assumption on φ leads (for neighboring points u and $u_0 \in E$) to the following chain of inequalities:

$$
\begin{aligned}
\|\overline{U} - \overline{U}_0\|_{2m-1+\mu} &\le \|\overline{U} - \overline{U}_0\|_{2m+\mu} \\
&\le C \cdot \|L(\overline{U} - \overline{U}_0)\|_\mu \\
&= C \cdot \|\tilde{P}\varphi(\cdot, u, Du, \ldots, D^{2m-1}u) - \tilde{P}\varphi(\cdot, u_0, Du_0, \ldots, D^{2m-1}u_0)\|_\mu
\end{aligned}
\tag{5.9}
$$

from which the continuity of F follows immediately. Now from [Ni1], p. 47, and estimate (5.9) the compactness of F follows immediately.

Let us check that the operator equation $Fu = u$ is equivalent to (5.2). Suppose $u^{(0)} = (\bar{a}^{(0)}, \bar{u}^{(0)})$ is a solution of (5.2) and $\overline{A}^{(0)} = A(u^{(0)})$, $\overline{U}^{(0)} = Q(u^{(0)})$. Then

$$
\varphi(\cdot, u^{(0)}, Du^{(0)}, \ldots, D^{2m-1}u^{(0)}) = Lu^{(0)} \in \tilde{E}
$$

from which (see (5.7))

$$
\langle \varphi(\cdot, u^{(0)}, Du^{(0)}, \ldots, D^{2m-1}u^{(0)}), w_i \rangle = 0 ,
$$

i.e. $\overline{A}^{(0)} = \overline{a}_i^{(0)}$ for all $i = 1, \ldots, k$. In addition,

$$\varphi(\cdot, u^{(0)}, Du^{(0)}, \ldots, D^{2m-1}u^{(0)}) = Lu^{(0)} = L(\overline{a}^{(0)}, \overline{u}^{(0)})$$
$$= L\overline{u}^{(0)} = \check{P}\varphi(\cdot, u^{(0)}, Du^{(0)}, \ldots, D^{2m-1}u^{(0)})$$
$$= L\overline{U}^{(0)} \ .$$

Since L is an injective operator on the complement to $\ker P$ we obtain $\overline{u}^{(0)} = \overline{U}^{(0)}$. Thus

$$Fu^{(0)} = F(\overline{a}^{(0)}, \overline{u}^{(0)}) = (\overline{a}^{(0)}, \overline{u}^{(0)}) = u^{(0)} \ .$$

Conversely, if $u^{(0)}$ is a fixed point for F, then $\overline{A}_i^{(0)} = \overline{a}_i^{(0)}$ $(i = 1, \ldots, k)$, i.e. (see (5.5) and (5.7))

$$\varphi(\cdot, u^{(0)}, Du^{(0)}, \ldots, D^{2m-1}u^{(0)}) = \check{P}\varphi(\cdot, u^{(0)}, Du^{(0)}, \ldots, D^{2m-1}u^{(0)})$$
$$= L\overline{U}^{(0)} = L(\overline{a}^{(0)}, \overline{u}^{(0)}) = Lu^{(0)} \ .$$

To complete the proof denote by S_r the sphere in E with radius equal to r and center coinciding with the origin. Estimate (5.4) provides the existence of $r_0 \geq 0$ (see [Ni1], p. 49), such that for all $r \geq r_0$ the vector field $(I - F)|E^G \cap S_r$ is non-degenerate and homotopic to I.

Now suppose that for some $r_1 \geq r_0$ problem (5.2) does not have a solution on the sphere S_{r_1}. This means that the vector field $I - F$ is non-degenerate on S_{r_1} and thus the winding number $\gamma(I - F, S_{r_1}, 0)$ is correctly defined for it. Then on the one hand $\gamma(I - F, S_{r_1}, 0) \neq 0$ according to Theorem 4.2; on the other hand an image of $I - F$ has a non-zero co-dimension (since $k > d$) and hence (see [Ni1], [Ra1]) $\gamma(I - F, S_{r_1}, 0) = 0$, which is a contradiction.

Using the standard regularity arguments (see, for instance, [Ni1], p. 50) one can easily prove that the obtained solution belongs to $C^\infty(\overline{\Omega}, R^q)$. This completes the proof of Theorem 5.1.

5.1.4. **Example 5.1.** Here we construct an example illustrating our result.

Let Ω be an open unit disk on the plane and $x = (x_1, x_2)$. Consider the system

$$\Delta u_1(x) = \exp\left(-\left(u_1(x) + \frac{\partial u_1}{\partial x_1} + \frac{\partial u_1}{\partial x_2}\right)^2 + \right.$$
$$\left. + u_2^2(x) + \left(\frac{\partial u_2}{\partial x_1}\right)^2 + \left(\frac{\partial u_2}{\partial x_2}\right)^2\right), \quad x \in \Omega \ , \tag{5.10}$$
$$\Delta u_2(x) = u_2(x) \cdot \exp(u_1(x)), \quad x \in \Omega \ ,$$
$$u_1(x) = 0, \quad x \in \partial\Omega \ ,$$
$$\frac{\partial u_2}{\partial x_1} = 0, \quad x \in \partial\Omega$$

and show that it satisfies the conditions of Theorem 5.1 (here Δ is the Laplacian).

Define an action of the group Z_2 on the space R^2 – the range of values of the vector function $u = (u_1, u_2)$ – by the formula: $g(u_1, u_2) = (u_1, -u_2)$.

The equivariance of the operator B which is defined by the boundary conditions (5.10) is obvious since $\partial(-u_2)/\partial x_1 = -\partial u_2/\partial x_1$ and g does not change the coordinate u_1. It is clear that the linear operator L, which is defined with the help of the Laplacians, is also equivariant. Let us convince ourselves that φ is equivariant. We have:

$$\varphi(g(u_1, u_2)) = \varphi(u_1, -u_2)$$

$$= \left(e^{-\left(u_1 + \frac{\partial u_1}{\partial x_1} + \frac{\partial u_1}{\partial x_2}\right)^2 + u_2^2 + \left(\frac{\partial u_2}{\partial x_1}\right)^2 + \left(\frac{\partial u_2}{\partial x_2}\right)^2}, -u_2 \cdot e^{u_1} \right)$$

$$= g\varphi(u_1, u_2) \ .$$

Now we will calculate the index of the linear problem associated with (5.10). It is clear that the desired index is equal to the sum of the indices of the Dirichlet problem and the problem with the direct derivative. It is well-known that the first of them is equal to zero (see, for example, [Hör], Section 10.5); the second is equal to $2 - 2\delta$, where δ is the degree of the map induced by the direct derivative (see [Hör], p. 266). In our situation condition $\partial u_2/\partial x_1 = 0$ implies $\delta = 0$. Thus the index we are interested in is equal to $2 > 0$.

It should be noticed that the co-kernel of the linear operator which is associated with (5.10) is trivial.

Let us verify estimate (5.4). The set E^G consists of functions presented in the form $(u_1(x), 0)$. So the corresponding restriction of φ has the following form:

$$\varphi = \exp\left(-\left(u_1 + \frac{\partial u_1}{\partial x_1} + \frac{\partial u_1}{\partial x_2} \right)^2 \right) \ .$$

Since

$$\left\| \exp\left(-\left(u_1 + \frac{\partial u_1}{\partial x_1} + \frac{\partial u_1}{\partial x_2} \right)^2 \right) \right\| < 1 \cdot \left(1 + \left\| \frac{\partial u_1}{\partial x_1} \right\| + \left\| \frac{\partial u_1}{\partial x_2} \right\| \right)^{1/2}$$

we see that (5.4) is true.

It remains to note that the invertibility of the operator $P|E^G$ is obvious because we are dealing with the Dirichlet problem.

5.1.5. Remark 5.1.   It is clear that if $u^{(0)}$ is a solution of (5.2), then $gu^{(0)}$ is also a solution for every $g \in G$. Therefore, under the conditions of Theorem 5.1 one can provide at least p solutions of (5.2) on the corresponding spheres, where p is the minimal "length" of the non-trivial orbits.

Remark 5.2. We have established our result under the assumption that estimate (5.4) holds. But from the proof it follows that it is suffices to require that $(I - F)|E^G \cap S_r$ is homotopic to the identical field for arbitrarily large r. However, this condition is harder to verify than estimate (5.4).

5.2. Genus of some subsets of G-spheres

5.2.1. A problem of estimating the genus of G-spaces (G-category, G-index, cup-length etc.) is attracting a big deal of attention (see, for instance, the recent monographs [Bar1, MW, Ra2, Str]) because of its great importance in minimax methods. At least two approaches to this problem exist: a geometric one based on Borsuk-Ulam type theorems and a homological one based on (co)homological arguments in the study of orbit spaces.

In this section we estimate the genus of a free part of a finite-dimensional sphere on which an arbitrary compact Lie group action is given. With Corollary 2.1 in hands we follow the geometric approach. The obtained result is applied to the situation of a unit sphere in the space of spherical harmonics under the natural representation of the group $SO(n)$. The study of the aforementioned situation becomes natural due to investigations of a bifurcation phenomenon in semi-linear elliptic equations on a ball (see, for instance, [Bar1]).

Let G be a compact Lie group which acts freely on a metric space M. By Lemma 1.1 the orbit map $p : M \to M/G$ is the projection of a locally trivial fiber bundle with fiber G.

Definition (see [Šv], p.250). *The minimal cardinality of an open covering of M/G consisting of the sets over which the fiber bundle is trivial, is said to be the genus of M (denoted by $\mathrm{gen}(M)$).*

Theorem 5.2. *Let G be a compact Lie group of dimension m acting smoothly on the sphere S^n. Let A be a closed G-invariant subset of S^n such that the G-space $S^n \setminus A$ is free. Suppose, further, A is an image of a k-dimensional smooth compact manifold under a smooth map with $k < n$ (if A is empty then it is thought as the image of the (-1)-dimensional manifold under the empty map). Then*

$$\mathrm{gen}(S^n \setminus A) \geq \frac{n - k}{1 + m}.$$

By the well-known properties of the Lusternik-Schnirelman category one has

Corollary 5.1.   *Under the conditions of Theorem 5.2*

$$\text{cat}((S^n \setminus A)/G) \geq \frac{n-k}{1+m}$$

Remark 5.3.   The smoothness condition on A in Theorem 5.2 does not seem to be so restrictive. Indeed, the union of all non-principal orbits does satisfy this condition.

Remark 5.4.   Our proof of Theorem 5.2 follows a geometric scheme by M. Krasnoselskii [Kr1, KZ]. The main ingredients of Krasnoselskii's investigation of a free case are the following : (a) the usage of the fact that the degree of any equivariant mapping of S^n into itself is different from zero, and (b) the observation that some simple operations increase dimension of subsets of a sphere not more than by one; this in turn allows him to use an induction over dimension. We also use an induction over dimension (Lemma 5.2), but the non-free situation we are dealing with forces us to supply the considered action and subsets with some additional structure. Instead of (a) we use a corresponding assertion for non-free actions (Corollary 2.1).

In the sequel we essentially use the following standard construction.

Definition.   For any $B \subset S^n$, $B \neq \emptyset$, and any $x_0 \in S^n$ with $-x_0 \notin B$ define a *(spherical) cone* over B with the vertex x_0 by the formula

$$\text{Con}(x_0, B) = \varphi^{-1}(\{(1-t)\varphi(x) + t\varphi(x_0)| \ x \in B, \ t \in [0,1]\}),$$

where φ is the stereographic projection of $S^n \setminus \{-x_0\}$ on R^n. Set also $\text{Con}(x_0, \emptyset) = \{x_0\}$.

Proposition 5.1.   Let B be a compact set in S^n and $x_0 \notin B$. Then for any neighborhood $U \supset \text{Con}(x_0, B)$ there exists a compact contractible and locally contractible set K such that $K \subset U, B \subset \text{Int}K$ and $\text{Con}(x_0, B) \subset K$.

This result is a direct consequence of the previous definition.

5.2.2.   The proof of Theorem 5.2 is based on the following lemma.

Lemma 5.2.   *Under the conditions of Theorem 5.2 there exists a chain of compact sets*

$$A \subset A_0 \subset B_0 \subset A_1 \subset B_1 \subset \ldots \subset B_{p-1} \subset A_p \subset S^n \qquad (5.11)$$

with the following properties:

(a) $A_p \neq S^n$;

(b) $p = \lceil (n-k)/(1+m) \rceil$, where $\lceil x \rceil = \min\{a \in Z | \ a \geq x\}$;

(c) B_i *is contractible and locally contractible;*

(d) A_{i+1} *is the union of all G-orbits passing through the points of B_i;*

(e) A_0 *is an invariant neighborhood of A*

$(i = 0, 1, ..., p-1)$.

Proof of Lemma 5.2. In order to prove the lemma construct a chain of compact subsets

$$A = \tilde{A}_0 \subset \tilde{B}_0 \subset \tilde{A}_1 \subset \tilde{B}_1 \subset \ldots \subset \tilde{B}_{p-1} \subset \tilde{A}_p \subset S^n \qquad (5.12)$$

satisfying the following properties:

(α) $\tilde{A}_p \neq S^n$;

(β) $\tilde{B}_i$ is the cone over $\tilde{A}_i$ with a vertex $x_i \notin \tilde{A}_i$;

(γ) $\tilde{A}_{i+1}$ is the union of all G-orbits passing through the points of $\tilde{B}_i$;

(δ) $\mu(\tilde{A}_{i+1}) = \mu(\tilde{B}_i) = 0$, where $i = 0, 1, \ldots, p-1$, and $\mu(\cdot)$ is the Lebesgue measure.

To this end we construct simultaneously by induction two chains of sets

$$A = A_0' \subset B_0' \subset A_1' \subset B_1' \subset \ldots \subset B_{p-1}' \subset A_p' \subset S^n \qquad (5.12')$$

$$\hat{A}_0 \subset \hat{B}_0 \subset \hat{A}_1 \subset \hat{B}_1 \subset \ldots \subset \hat{B}_{p-1} \subset \hat{A}_p$$

and a chain of smooth maps

$$f_0 \subset g_0 \subset f_1 \subset g_1 \subset \ldots \subset g_{p-1} \subset f_p$$

such that chain $(5.12')$ satisfies properties $(\alpha) - (\delta)$ above (with A_i', B_i' instead of $\tilde{A}_i$, $\tilde{B}_i$ respectively) and

$f_0 : \hat{A}_0 \to A_0' = \tilde{A}_0 = A$ is the map defined by the lemma assumptions;

$f_{i+1}(\hat{A}_{i+1}) = A_{i+1}'$;

$g_i(\hat{B}_i) = B_i'$;

$\dim \hat{A}_0 = k$, $\dim \hat{A}_i = k + (m+1)i$, $\dim \hat{B}_i = k + (m+1)i + 1$,

where $i = 0, 1, ..., p-1$.

By the assumptions the smooth surjective map $f_0 : \hat{A}_0 \to A = A_0' = \tilde{A}_0$ is given. Suppose, a smooth map $f_i : \hat{A}_i \to A_i'$ has been constructed. Set $\hat{B}_i = \hat{A}_i \times R$

for $i = 1, ..., p - 1$. By the inductive hypothesis A'_i is of zero Lebesgue measure in S^n. Hence, there exists a point $x_i \in S^n$ such that $x_i, -x_i \notin A'_i$. Let us consider the stereographic projection $\varphi : S^n \setminus \{-x_0\} \to R^n$ and set

$$g_i(x, t) = \varphi^{-1}((1 - t)\varphi(f_i(x)) + t\varphi(x_i)), \qquad B'_i = g_i(\hat{B}_i)$$

for all $x \in \hat{A}_i$ and $t \in R$. If $A = \emptyset$ we suppose $\hat{B}_0$ is a point and the map $g_0 : \hat{B}_0 \to S^n$ is a constant map to the vertex of the cone over the empty set. Since $\dim A'_i = k + (m + 1)i$ then $\dim B'_i = k + (m + 1)i + 1 < n$ and (according to the Sard Theorem) $\mu(B'_i) = 0$. The smoothness of g_i is obvious.

Set $\hat{A}_{i+1} = G \times \hat{B}_i$ and for every $g \in G$ and every $x \in \hat{B}_i$ define $f_{i+1}(g, x) = gx$. Clearly, f_{i+1} is a smooth map. Since $\dim \hat{A}_{i+1} = k + (m + 1)(i + 1) < n$ then (by the Sard Theorem) $\mu(A'_{i+1}) = 0$. Thus chain (5.12') has been constructed.

Chain (5.12') satisfies all the required properties of chain (5.12) but compactness. Set (for all $i = 0, ..., p - 1$)

$$\tilde{A}_0 = A;$$

$$\tilde{B}_i = g_i(f_i^{-1}(A'_i) \times [0, 1]) \subset B'_i;$$

$$\tilde{A}_{i+1} = f_{i+1}(G \times g_i^{-1}(\tilde{B}_i)) \subset A'_{i+1}.$$

By construction, $\tilde{B}_i = \mathrm{Con}(x_i, \tilde{A}_i)$ and $\tilde{A}_{i+1} = G(\tilde{B}_i)$. Compactness of $\tilde{A}_i$ and $\tilde{B}_i$ could be established by induction. It should be noticed that $\tilde{A}_p \neq S^n$ since $\mu(\tilde{A}_p) = 0$.

We produce chain (5.11) from chain (5.12) . Since $\tilde{A}_p$ is a proper invariant compact subset of S^n there exists a closed invariant neighborhood $\bar{A}_p$ of $\tilde{A}_p$ such that $\bar{A}_p \neq S^n$. Since $\hat{A}_p$ is a neighborhood of $\tilde{B}_{p-1}$ there exists a compact contractible and locally contractible set B_{p-1} such that $\tilde{B}_{p-1} \subset B_{p-1} \subset \bar{A}_p$ (see Proposition 5.1). Bearing in mind the invariance of $\bar{A}_p$ we see that $G(B_{p-1}) \subset \bar{A}_p$; set $A_p = G(B_{p-1})$. According to Proposition 5.1, $\tilde{A}_{p-1} \subset \mathrm{Int} B_{p-1}$, hence by invariance of $\tilde{A}_{p-1}$ there exists an invariant closed neighborhood $\bar{A}_{p-1}$ of set $\tilde{A}_{p-1}$ such that $\bar{A}_{p-1} \subset B_{p-1}$.

Now applying the described procedure "down" one can construct the required sets $A_{p-1}, ..., B_0, A_0$.

Lemma 5.2 is proved.

5.2.3.   Proof of Theorem 5.2. To prove the theorem it suffices to show the existence of an invariant compact $K \subset S^n \setminus A$ with $\mathrm{gen}(K) \geq p$. Let us suppose this assertion to be false, i.e. for any invariant compact set $K \subset S^n \setminus A$ $\mathrm{gen}(K) < p$. Consider chain (5.11) from Lemma 5.2 and let $K \subset S^n \setminus A$ be an invariant compactum such that $A_0 \cup K = S^n$. By the assumption $\mathrm{gen}(K) = \ell < p$. Hence there exist invariant open (in the induced topology) subsets $M_1, \ldots, M_\ell$ of K such that $\bigcup_{i=1}^{\ell} M_i = K$ and $\mathrm{gen}(M_i) = 1$.

Consider the chain of closed invariant sets

$$A_0 = K_0 \subset K_1 \subset \ldots \subset K_\ell = S^n,$$

where $K_i = (S^n \setminus (M_{i+1} \cup \ldots \cup M_\ell)) \cup A_0$ for $0 \le i \le \ell$. Note that for $i \ge 1$ one has $P_i = K_{i+1} \setminus K_i \subset M_i$ from which it follows that the projection of P_i on K/G defines a trivial fiber bundle. Hence for any $i \ge 1$ there exists a compact set $L_i \subset K$ such that $P_i = G(L_i)$ and $g(L_i) \cap h(L_i) = \emptyset$ if $g \ne h$ $(g, h \in G)$.

Now we construct an equivariant map $\Phi : S^n \to S^n$ such that $\Phi|A$ is the identity map and $\Phi(S^n) \ne S^n$; this contradicts Corollary 2.1.

Set Φ to be the identity map on A_0. The construction of Φ is proceeded by induction. If the map Φ is constructed on K_i, and its image belongs to A_i then we may extend Φ to a continuous (non-equivariant) map $L_{i+1} \to B_i$ (this is possible since K_i is closed and A_i is an AR-space. After that we extend Φ over all of the sets M_i by $\Phi gx = g\Phi x$ $(x \in L_i, g \in G)$. The image of the final map is contained in $A_\ell \subset A_p \subset S^n$ and $A_p \ne S^n$.

Theorem 5.2 is proved.

5.2.4.   In this subsection we apply Theorem 5.2 to the situation of an action of a group $SO(n)$ (n is odd) on the unit sphere in a space of spherical harmonics. Denote by $P(n, \ell)$ the space of all homogeneous polynomials of degree ℓ in n variables and by $H(n, \ell)$ – the corresponding space of spherical harmonics ($H(n, \ell) \subset P(n, \ell)$). Let $x, y_1, z_1, \ldots, y_m, z_m$ be an orthogonal basis in space R^n, $n = 2m + 1$. It is well known (see, for instance, [BtD]) that a polynomial $f \in P(n, \ell)$ of the form

$$f = \sum_{k=0}^{\ell} \frac{x^k}{k!} f_k(y_1, z_1, \ldots, y_m, z_m) \tag{5.13}$$

belongs to $H(n, \ell)$ iff

$$\forall k : \ 0 \le k \le \ell - 2 : \ f_{k+2} = -\Delta f_k, \tag{5.14}$$

$H(n, \ell) = H_0 \oplus H_1$, where H_0 consists of the harmonics containing even degrees of x and H_1 consists of the harmonics containing odd degrees of x. Moreover, a polynomial f from H_0 or H_1 is uniquely determined by its leading term f_0 or f_1 correspondingly. It is well known that $H_0 \cong P(n-1, \ell)$, $H_1 \cong P(n-1, \ell-1)$ and

$$\dim P(n, \ell) = \binom{n + \ell - 1}{\ell},$$

$$\dim H(n, \ell) = \dim P(n - 1, \ell) + \dim P(n - 1, \ell - 1).$$

The expression $(gf)(u) = f(g^{-1}(u))$ defines the standard (irreducible) representation of $SO(n)$ on $H(n, \ell)$.

In order to apply Theorem 5.2 one has to calculate

$$d(n, \ell) = \max\{\dim H(n, \ell)^g \mid g \in SO(n), g \neq 1\},$$

where $H(n, \ell)^g = \{f \in H(n, \ell) \mid gf = f\}$.

Let g be a non-trivial element of $SO(n)$ and let $x, y_1, z_1, \ldots, y_m, z_m$ be an orthonormal basis in R^n where g has the following form:

$$\begin{pmatrix} 1 & & & 0 \\ & A(\varphi_1) & & \\ & & \ddots & \\ 0 & & & A(\varphi_m) \end{pmatrix}$$

where $A(\varphi_k)$ is the rotation matrix by the angle φ_k $(1 \leq k \leq m)$. Set $x = x$, $\bar{y}_k = y_k + iz_k$, $\bar{z}_k = y_k - iz_k$. In the new basis g is presented as a diagonal matrix

$$\mathrm{diag}(1, \exp(i\varphi_1), \exp(-i\varphi_1), \ldots, \exp(i\varphi_m), \exp(-i\varphi_m)). \tag{5.15}$$

It is easy to see that

$$\Delta f(\bar{y}_1, \bar{z}_1, \ldots, \bar{y}_m, \bar{z}_m) = 4 \sum_{k=1}^{m} \frac{\partial^2 f}{\partial \bar{y}_k \partial \bar{z}_k},$$

and consequently (in accordance with (5.13–5.14))

$$f_{k+2} = -4 \sum_{k=1}^{m} \frac{\partial^2 f_k}{\partial \bar{y}_k \partial \bar{z}_k}. \tag{5.16}$$

Set $B(m, \ell) = B^{(0)}(m, \ell) \cup B^{(1)}(m, \ell)$, where $B^{(r)}(m, \ell)$ is the set of all non-negative integer-valued $2m$-vectors such that the sum of its coordinates equals $\ell - r$ $(r = 0, 1)$. Given a vector b from $B^{(r)}(m, \ell)$ one can consider a polynomial (uniquely determined by b) $f^{(b)} \in H(n, \ell)$ with the leading term

$$x^r \bar{y}_1^{b(1)} \bar{z}_1^{b(2)} \ldots \bar{y}_m^{b(2m-1)} \bar{z}_m^{b(2m)}.$$

The family of all these polynomials forms a basis of the space $H(n, \ell)$, and (5.13)–(5.16) show the diagonality of the action of g in this one:

$$gf^{(b)} = \exp(i \sum_{k=1}^{m} \varphi_k(b(2k - 1) - b(2k))) f^{(b)}.$$

Hence dim $H(n, \ell)^g$ is equal to the number of the g-fixed polynomials from our basis, i.e. it is equal to the number of vectors $b \in B(m, \ell)$ such that

$$\sum_{k=1}^{m} t_k s_k^{(b)} \in Z, \tag{5.17}$$

where $t_k = \varphi_k / 2\pi$ and $s_k^{(b)} = b(2k-1) - b(2k)$. In what follows, we vary g in order to maximize dim $H(n, \ell)^g$ so that the matrix of g preserves form (5.15) in the basis described above. The group of all these matrices we denote by Λ. Without loss of generality, suppose the numbers t_k from (5.17) to be rational: $t_k = u_k / v_k$ $(u_k, v_k \in Z)$, $t_k \neq 0$ if $1 \leq k \leq \alpha$ and $t_k = 0$ if $k > \alpha$ for some integer α $(1 \leq \alpha \leq m)$. Thus, dim $H(n, \ell)^g$ is equal to the number of vectors from $B(m, \ell)$ satisfying the congruence

$$\sum_{k=1}^{\alpha} u_k v_k' s_k^{(b)} \equiv 0 \pmod{v}, \tag{5.18}$$

where $v = \mathrm{lcm}(v_1, \ldots, v_\alpha)$ and v_k' is the corresponding complementary multiplier. Obviously, every vector b satisfying congruence (5.18) also satisfies the congruence

$$\sum_{k=1}^{\alpha} a_k s_k^{(b)} \equiv 0 \pmod{p}, \tag{5.19}$$

where p is a prime divisor of v, all $a_k \neq 0$ and $a_1 = 1$ (maybe for some smaller value of α). Fixing all the components of the vector b but the first one, it is easy to verify that the number of solutions of (5.19) is maximal for $p = 2$; denote this number by $c(m, \ell, \alpha)$. Simple combinatorial arguments show that

$$c(m, \ell, \alpha) = \begin{cases} \sum_{k=0}^{[(\ell-1)/2]} \dim H(2m - 2\alpha, \ell - 2k), & \text{if } \alpha < m; \\ \dim P(2\alpha, 2[(\ell-1)/2]), & \text{if } \alpha = m, \end{cases} \tag{5.20}$$

where $[\cdot]$ means the integer part of a number.

Let us introduce the following notations: for $\alpha \in [1, m] \cap Z$

$$B_r^\alpha = B_r^\alpha(m, \ell) = \{b \in B(m, \ell) \mid \sum_{k=1}^{\alpha} s_k^{(b)} \equiv r \pmod{2}\},$$

and, for any $a \in B(m, \ell)$,

$$B_a = B_a(m, \ell) = \{b \in B(m, \ell) \mid \forall\, 3 \leq k \leq 2m - 2 : b(k) = a(k)\}.$$

Let ℓ be an even number. Let us show that for any $\alpha = 1, 2, \ldots, m-1$, $c(m, \ell, \alpha) \leq c(m, \ell, m)$. Consider B_0^α as the union of two disjoint sets $B_0^\alpha \cap B_0^m$ and

$B_0^\alpha \cap B_1^m$ and define a mapping $\varphi : B_0^\alpha \to B_0^m$ by setting $\varphi(b) = b \in B_0^\alpha \cap B_0^m$ for $b \in B_0^\alpha \cap B_0^m$ and $\varphi(b) = b + (1, 0, \ldots, 0) \in B_1^\alpha \cap B_0^m$ for $b \in B_0^\alpha \cap B_1^m$. Now the required inequality follows from injectivity of φ.

Similarly, one can prove that $c(m, \ell, \alpha) \geq c(m, \ell, m)$ for any $\alpha = 1, \ldots, m - 1$ if ℓ is odd.

In what follows, we will show that for any ℓ and for all $\alpha = 2, \ldots, m - 1$ the following inequality holds: $c(m, \ell, \alpha) \leq c(m, \ell, 1)$. Let us represent B_0^α as the union of the disjoint sets $B_0^\alpha \cap B_0^1$ and $B_0^\alpha \cap B_1^1$. The second of these two sets can be represented as the disjoint union of the family $\Gamma(m, \ell) = \{B_b(m, \ell) \mid b \in B(m, \ell)\}$. For fixed b every vector from $B_b(m, \ell)$ is determined uniquely by its four coordinates with indices 1, 2, $2m - 1$ and $2m$. The direct calculation according to (5.20) shows that $\mid B_0^1(2, k) \mid > \mid B_1^1(2, k) \mid$. Hence there exists an injective mapping $\varphi_b : B_b \cap B_1^1 \to B_b \cap B_0^1$ such that $\varphi_b(B_0^\alpha \cap B_1^1 \cap B_b) \subset B_1^\alpha \cap B_0^1 \cap B_b$. Taking the union of the family $\{\varphi_b \mid b \in B(m, l)\}$ and the identity map on $B_0^\alpha \cap B_0^1$ one can obtain the injection required: $\varphi : B_0^\alpha \to B_0^1$.

The arguments above show that the number $c(m, \ell, \alpha)$ is maximal for $\alpha = 1$ if ℓ is odd, and for $\alpha = m$, if ℓ is even. Hence (see (5.20)) we obtain the following formulas:

$$d(n, \ell) = \begin{cases} \sum_{k=0}^{(\ell-1)/2} \dim H(2m - 2, \ell - 2k), & \text{if } \ell \text{ is odd}; \\ \dim P(2m, \ell) & \text{if } \ell \text{ is even.} \end{cases} \qquad (5.21)$$

Using the natural multi-dimensional generalization of the Euler angles one can construct a smooth mapping φ from the torus T^{2m^2} to $SO(n)$ such that any matrix $P \in SO(n)$ can be represented in the form $P = LDL^{-1}$, where $L \in \mathrm{Im}\varphi$ and $D \in \Lambda$, thus $SO(n)(H(n, \ell)^g) = (\mathrm{Im}\varphi)(H(n, \ell)^g)$. Now, if S is the unit sphere in $H(n, \ell)$, A is the union of all non-principal orbits, then A coincides with the union of the family $\bar{A}(\Lambda) = \{(\mathrm{Im}\varphi)(H(n, \ell)^g) \mid g \in \Lambda\}$ and (according to the finiteness of the number of orbit types of $SO(n)$-action on S, see [Bre]) is equal to the union of some finite family $\bar{A}(\lambda)$, where λ is a finite subset of Λ. Now, using the manifold

$$M = \bigsqcup_{g \in \lambda} (T^{2m^2} \times (H(n, \ell)^g \cap S) \times T^{d(g)}),$$

where $d(g) = d(n, \ell) - \dim H(n, \ell)^g$ and symbol $\bigsqcup$ means a disjoint union, define a surjection $f : M \to A$, as follows:

$$f(x, y, z) = \varphi(x)(y) \ ((x, y, z) \in M).$$

Applying Theorem 5.2 we get

Theorem 5.3.   *Let $n = 2m+1$ and let $H(n, \ell)$ be the space of spherical harmonics with the natural representation of the group $SO(n)$. Let, further, S be the unit sphere in $H(n, \ell)$, and $A \subset S$ be the union of all non-principal orbits. Then*

$$\text{gen}(S \setminus A) \geq \lceil \gamma(n, \ell) \rceil,$$

where

$$\gamma(n, \ell) = \frac{\dim H(n, \ell) - d(n, \ell) - 2m^2}{m(2m + 1) + 1},$$

and $d(n, \ell)$ is defined by (5.21).

Finally, let us discuss some properties of the function $\gamma(n, \ell)$.

1) $\gamma(3, \ell) = [(\ell + 1)/4]$.

2) Let $n \geq 3$. For $\ell = 1, 2$, the set A coincides with the entire sphere S; $\gamma(n, 1) = \gamma(n, 2) = 0$. One can show that $\gamma(n, 3) = 1$ and $\gamma(n, 4) = m - [m/3]$.

3) Using simple arguments one can show that for fixed n $\gamma(n, \ell)$ does not decrease in ℓ . Although calculations confirm the conjecture that the function $\gamma(n, \ell)$ never decreases, we have no proof of this statement.

5.3. Equivariant Hopf Theorem

5.3.1.   Let us recall a classical theorem of H. Hopf (see, for instance, [Di2], p. 122). Let M^n be a closed, compact, connected, oriented n-dimensional manifold. Choose generators $z(S^n)$ and $z(M^n)$ of $H^n(S^n; Z)$ and $H^n(M^n; Z)$. Given $f : M^n \to S^n$ its degree $\deg f \in Z$ is defined by $f^* z(S^n) = (\deg f) \cdot z(M^n)$ (as is well-known (see, for instance, [Sp]), this definition is equivalent to the one given in Chapter 1). Hopf's theorem then asserts that the correspondence $[f] \to \deg f$ is a bijection between homotopy classes of maps from M^n to S^n and Z. Suppose now that a compact Lie group G acts on M^n and S^n. Classification of equivariant maps $M^n \to S^n$ up to equivariant homotopy cannot be achieved in the same straightforward way as in the non-equivariant case. As an example, suppose that G is a finite group acting orthogonally on vector spaces V and W. Suppose that for all subgroups $H \subset G$ the dimensions of fixed point sets V^H and W^H are equal. Take the unit spheres $S(V)$ and $S(W)$ and consider the following statement :

(∗) *G-equivariant maps $f_1, f_2 : S(V) \to S(W)$ are equivariantly homotopic if and only if*

$$\deg(f_1|S(V)^H) = \deg(f_2|S(W)^H)$$

for all subgroups H of G.

Although this statement is not true in general (see [Ru]) , there exists a rather general set of conditions on G-spaces V and W which ensure its validity. These conditions can be obtained as a corollary of the so called Equivariant Hopf Theorem presented by   tom Dieck in [Di1,Di2] and generalized by J. Tornehave [To] and E. Laitinen [Lai]. In this section we discuss a more straightforward approach to the Equivariant Hopf Theorem which enables us to obtain *necessary and sufficient* conditions for statement ($*$). We also generalize results on equivariant homotopy classification obtained in [Di1, Di2, To, Lai].

5.3.2.   Hereafter, all the cohomology groups are considered over the integers Z.

Let (K, L) be a relative CW-complex, $\dim(K \backslash L) \leq n$ $(n \geq 1)$. Suppose, further, that the metric space Y is locally and globally k-connected for all $k = 0, 1, \ldots, n-1$, $\pi_n(Y) = Z$ and O is a generator of the group $H^n(Y) = Z$. Then the following two results are well-known.

Proposition 5.2   (see, for instance, [Hu], Theorem 15.1, p. 191; Lemma 16.1, p. 191; Theorem 16.3, p. 192). *Let (K, L) and Y be as above and let $f : L \to Y$ be a continuous map. Then the homotopic classes relative to L of the extensions of f over K are in one-to-one correspondence with the elements of the group $H^n(K, L)$. This correspondence is given in the following way. For a pair $\widetilde{f}, \widehat{f} : K \to Y$ of arbitrary extensions of f, there exists a unique element $\omega^n(\widetilde{f}, \widehat{f}) \in H^n(K, L)$ such that $\omega^n(\widetilde{f}, \widehat{f}) = 0$ iff $\widetilde{f}, \widehat{f}$ are homotopic relative to L, and the correspondence is determined by the assignment $\widetilde{f} \to \omega^n(\widetilde{f}, \widehat{f})$ for any given extension $\widehat{f}$ and all extensions $\widetilde{f}$ of f over K.*

Proposition 5.3   ([Hu], Proposition 13.2, p. 190). *Under the conditions of Proposition 5.2*

$$\widetilde{f}^*(O) - \widehat{f}^*(O) = j^*(\omega^n(\widetilde{f}, \widehat{f})) \,,$$

where j^* is the homomorphism from the cohomological exact sequence of the pair (K, L).

Now, in addition to the conditions of Proposition 5.2, assume that a finite group G acts on K and Y and that (K, L) is a relative free G-equivariant CW-complex (see [Di2], p. 98). In this case $H^n(K, L)$ is a ZG-module. The action of G on $H^n(Y) = Z$ induces a homomorphism $e_{G,Y} : G \to \{-1, 1\}$ (cf. Remark 1.1). Set

$$w = \sum_{g \in G} e_{G.Y}(g^{-1}) \cdot g \,.$$

It is clear that $w \cdot H^n(K, L)$ is a ZG-submodule of the module $H^n(K, L)$. The following fact is the starting point of our discussion.

Basic Lemma.   *Let (K, L) and Y be as above and let $f : L \to Y$ be an equivariant map. Denote by K_{n-1} the G-invariant $(n-1)$-skeleton of K.*

(i) If $\widehat{f}, \widetilde{f} : K \to Y$ is a pair of equivariant extensions of f then the element $\omega^n(\widehat{f}, \widetilde{f})$ from Proposition 5.2 belongs to the ZG-submodule $w \cdot H^n(K, L)$.

(ii) For any $w \cdot \alpha \in w \cdot H^n(K, L)$ and any equivariant extension $\widehat{f} : K \to Y$ of f, there exists an equivariant extension $\widetilde{f} : K \to Y$ of f such that $\omega^n(\widehat{f}, \widetilde{f}) = w \cdot \alpha$.

(iii) If $\widehat{f}, \widetilde{f} : K \to Y$ is a pair of equivariant extensions of f, then an equivariant homotopy between $\widehat{f}$ and $\widetilde{f}$ on $L \cup K_{n-1}$ equivariantly extends over K iff it has a (non-equivariant) extension over K.

Proof. (i) In accordance with Proposition 5.2, let us take an element $\omega^n(\widehat{f}, \widetilde{f}) \in H^n(K, L)$ and show that it has a special form.

By Corollary 1.1 (see also [Ja1, Di1, Di2]) there exists a G-equivariant homotopy between $\widehat{f}$ and $\widetilde{f}$ on the set $L \cup K_{n-1}$. This homotopy together with the maps $\widetilde{f}$ and $\widehat{f}$ defines a G-equivariant map

$$f_0 : (L \cup K_{n-1}) \times [0,1] \cup (K \times \{0,1\}) \to Y .$$

Since (K, L) is a relative free G-equivariant CW-complex we can construct a "cellular" fundamental domain D for the action of G on $K \backslash L$ picking up one cell from each "cellular" orbit of G. For any $(n+1)$-dimensional cell σ_r in $(K \backslash L) \times [0,1]$ we have

$$\partial \sigma_r \subset (L \cup K_{n-1}) \times [0,1] \cup (K \times \{0,1\}) .$$

For every σ_r define a cochain K_{σ_r} by the formula:

$$K_{\sigma_r} = \begin{cases} [f_0(\partial \sigma_r)] & \text{on } \sigma_r \\ 0 & \text{on the other cells} \end{cases}$$

(see [Hu], p. 176). Then

$$\nu = \sum_{\sigma_r \subset (K \backslash L) \times [0,1]} K_{\sigma_r} .$$

is a difference cochain (see [Hu], p. 178). Set

$$\nu_1 = \sum_{\sigma_r \subset D \times [0,1]} K_{\sigma_r}$$

and check that $\nu = w \cdot \nu_1$.

To this end suppose that σ_0 is a cell from $D \times [0,1]$. We will show that

$$w \cdot K_{\sigma_0} = \sum_{g \in G} K_{g(\sigma_0)} \ . \tag{5.22}$$

Let $\sigma_1 = g_1(\sigma_0)$ for some $g_1 \in G$. Then

$$(w \cdot K_{\sigma_0})(g_1)(\sigma_0) = ((\sum_{g \in G} e_{G,Y}(g^{-1}) \cdot g) \cdot K_{\sigma_0})(g_1(\sigma_0)) \tag{5.23}$$

$$= \sum_{g \in G} e_{G,Y}(g^{-1}) \cdot K_{\sigma_0}(gg_1(\sigma_0))$$

$$= e_{G,Y}(g_1^{-1}) \cdot K_{\sigma_0}(g_1^{-1}g_1(\sigma_0))$$

$$= e_{G,Y}(g_1^{-1}) \cdot K_{\sigma_0}(\sigma_0) \ .$$

On the other hand, taking into account equivariance of the map f_0 we have:

$$\sum_{g \in G} K_{g(\sigma_0)}(g_1(\sigma_0)) = K_{g_1(\sigma_0)}(g_1(\sigma_0)) \tag{5.24}$$

$$= [f_0(\partial(g_1(\sigma_0)))]$$

$$= [f_0(g_1(\partial(\sigma_0)))]$$

$$= [g_1^{-1}(f_0(\partial(\sigma_0)))]$$

$$= e_{G,Y}(g_1^{-1}) \cdot K_{\sigma_0}(\sigma_0) \ .$$

Now comparing (5.23) and (5.24) we obtain (5.22).

Thus

$$\nu = \sum_{\sigma_r \subset (K \backslash L) \times [0,1]} K_{\sigma_r}$$

$$= \sum_{\sigma_0 \subset D \times [0,1]} \sum_{g \in G} K_{g(\sigma_0)}$$

$$= \sum_{\sigma_0 \subset D \times [0,1]} w \cdot K_{\sigma_0}$$

$$= w \cdot \sum_{\sigma_0 \subset D \times [0,1]} K_{\sigma_0}$$

$$= w \cdot \nu_1 \ .$$

Using the one-to-one correspondence $\sigma_r \times [0,1] \to \sigma_r$ between the $(n+1)$-dimensional cells of the set $K \times [0,1] \backslash (L \times [0,1] \cup K \times \{0,1\})$ and the n-dimensional cells of the set $K \backslash L$ we obtain for the deformation cochain ([Hu], p. 179):

$$\tau(\nu) = \tau(w \cdot \nu_1)$$

$$= \tau(\sum_{g \in G} e_{G,Y}(g^{-1}) \cdot g) \cdot \nu_1$$

$$= (\sum_{g \in G} e_{G,Y}(g^{-1}) \cdot g)\tau(\nu_1)$$

$$= w \cdot \tau(\nu_1) \in w \cdot C^n(K,L) \ .$$

Finally, since the sets of cocycles and coboundaries are ZG-submodules of the ZG-module $C^n(K, L)$ one concludes that the cohomological class of the cocycle $\tau(\nu)$ has the form

$$\omega \cdot \alpha \, ,$$

for some $\alpha \in H^n(K, L)$.

(ii) Let $\widehat{f} : K \to Y$ be an arbitrary equivariant extension of f and let c be a cocyle which represents $w \cdot \alpha$. Without loss of generality one may suppose that $c = w \cdot \alpha'$, where $\mathrm{supp}(\alpha') \subset D$. From [Hu], Lemma 4.3, p. 179, it follows that there exists a (generally speaking, non-equivariant) map $\overline{f} : K \to Y$ such that $\overline{f}|L = f$ and $d^n(\widehat{f}, \overline{f}) = c$; by the construction of $d^n(\widehat{f}, \overline{f})$ one may assume that $\widehat{f}$ and $\overline{f}$ are equivariantly homotopic on $L \cup K_{n-1}$. Let us take restrictions $\widehat{f}|D$ and $\overline{f}|D$. It is clear that

$$\sum_{\sigma_r \in D} d^n_{\sigma_r}\left(\widehat{f}|D, \overline{f}|D\right) = \alpha' \, ,$$

where $d^n_{\sigma_r}\left(\widehat{f}|D, \overline{f}|D\right)$ is a difference cochain for $\widehat{f}|D$ and $\overline{f}|D$ on σ_r. Extend $\overline{f}|D$ to the equivariant map $\widetilde{f} : K \to Y$ in accordance with Lemma 1.16. Obviously $w^n(\widehat{f}, \widetilde{f}) = w \cdot \alpha'$. This completes the proof of (ii).

(iii) In order to prove the last statement it suffices to take a restriction of the (non-equivariant) homotopy to the fundamental domain and to extend it by equivariance.

Corollary 5.2.   *Under the assumptions of Basic Lemma, the equivariant homotopic classes relative to L of the equivariant extensions of f over K are in one-to-one correspondence with the elements of the submodule $w \cdot H^n(K, L)$ of $H^n(K, L)$. The correspondence is determined by the assignment $\widetilde{f} \to w^n(\widehat{f}, \widetilde{f})$ for any given equivariant extension $\widehat{f}$ and all equivariant extensions $\widetilde{f}$ of f over K.*

5.3.3.   We now formulate a general version of the Equivariant Hopf Theorem.

We make the following assumptions (see [Di1], p. 212; [Di2], p. 125, [To], [Lai]).

(i) G is a compact Lie group, X is a G-complex of a finite orbit type and Y is a G-space.

(ii) For every $H \in \mathrm{Iso}(X)$ the space X^H is a finite-dimensional $W(H)$-complex of topological dimension $n(H) > 0$; the space Y^H is locally and globally k-connected for each $k = 0, 1, \ldots, n(H) - \dim W(H)$, if $|W(H)| = \infty$ and locally and globally k-connected for each $k = 0, 1, \ldots, n(H) - 1$ otherwise.

(iii) Let $\Phi(G, X, Y) = \{H \in \mathrm{Iso}(X) : W(H) \text{ finite}, \pi_{n(H)}(Y^H) \neq 0\}$ (it is easy to see that $\Phi(G, X, Y)$ is closed under conjugations). For $H \in \Phi(G, X, Y)$

we assume that $\pi_{n(H)}(Y^H) = Z$; then $H^{n(H)}(Y^H) = Z$ and we fix the generators $O_H \in H^{n(H)}(Y^H)$.

As was mentioned above, for every $H \in \Phi(G, X, Y)$ the $ZW(H)$-module structure of $H^{n(H)}(Y^H)$ can be specified by a homomorphism $e_{W(H),Y} : W(H) \to \{+1, -1\}$ such that $\omega \in W(H)$ acts as a multiplication by $e_{W(H),Y}$. Let us denote the element

$$\sum_{h \in W(H)} e_{W(H),Y}(h^{-1}) \cdot h$$

by w_H.

Theorem 5.4. *Suppose that X and Y satisfy (i)-(iii). Then the following four statements hold.*

(a) The set $[X; Y]_G$ of the equivariant homotopy classes is not empty.

(b) Let $H \in \Phi(G, X, Y)$, $\dim(X^{>H}) < n(H)$ and $f_1^H, f_2^H : X^H \to Y^H$ are $W(H)$-equivariant maps such that $f_1^H | X^{>H}$ and $f_2^H | X^{>H}$ are $W(H)$-equivariantly homotopic. Then

$$(f_1^H)^*(O_H) \equiv (f_2^H)^*(O_H) \qquad (\mathrm{mod}\, w_H \cdot H^{n(H)}(X^H)) \ .$$

(c) Let $f : X^{>H} \to Y^H$ be a $W(H)$-equivariant map, let H be as in (b) and let $\beta \in H^{n(H)}(X^H)$. Then there exists a pair of $W(H)$-equivariant extensions $f_1^H, f_2^H : X^H \to Y^H$ of the map f such that

$$(f_1^H)^*(O_H) = (f_2^H)^*(O_H) + w_H \cdot \beta \ .$$

(d) The statement

"$f_1, f_2 : X \to Y$ are G-equivariantly homotopic if and only if $(f_1 | X^H)^*(O_H) = (f_2 | X^H)^*(O_H)$ for every $H \in \Phi(G, X, Y)$"

is equivalent to the following condition:

$$\mathrm{Ker}\, j^* | w_H \cdot H^{n(H)}(X^H, X^{>H}) = 0 \qquad (5.25)$$

for every $H \in \Phi(G, X, Y)$.

Remark 5.5. Note that one can reformulate Theorem 5.4 in a "relative form" compatible with Theorem 3.2′. In addition, it is easy to see that one can generalize the Basic Lemma to the case of arbitrary compact Lie groups. To this end it suffices to note that G acts on cohomology groups as G/G_0 where G_0 is the connected component of the unity, and to redefine the element w accordingly. We leave the corresponding reformulation of Theorem 5.4 to the reader.

Remark 5.6 (cf. [To, Lai]). Let X be a smooth (compact, closed, connected, oriented) G-manifold. Assume that X^H is oriented for all $H \in \mathrm{Iso}(X)$ and that $e_{W(H),X} = e_{W(H),Y}$. Then X satisfies condition (5.25) if the action of G on X has the following property :

for any H, $K \in \mathrm{Iso}(X)$, $H \subset K$, $H \in \Phi(G,X,Y)$ such that codimension of X^K in X^H is exactly one, the following condition is fulfilled:

$$N(H) \bigcap K \neq H. \qquad (5.25')$$

For example this property is obviously satisfied by any action of a nilpotent group (see [To, Lai]).

Proof. It follows from (5.25') that $W = W(H)$ acts transitively on the set of connected components of $X^H \setminus X^{>H}$ (it suffices to take a submanifold P with codim $P = 1$ which is fixed by $R = K \cap N(H)$, and consider the (non-trivial) action of R in a normal slice to P (see also [To])). Under the assumptions on X one can "represent" a component of $X^H \setminus X^{>H}$ by some $\tau \in H_{n(H)}(X^H, X^{>H})$. If $w_H \cdot \tau = 0$ then of course $w_H \cdot H^{n(H)}(X^H, X^{>H}) = 0$ and we are done. Hence we assume that $w_H \cdot \tau \neq 0$. Let σ be a generator of $H_{n(H)}(X^W)$. The transitivity of the action of W and the condition $e_{W,X} = e_{W,Y}$ yield the equality $j_*(\sigma) = (1/|W_\tau|)w_H \cdot \tau$. Denoting by $\delta \in H^{n(H)}(X^H, X^{>H})$ an element, which is dual to τ, we have

$$j^*(w_H \cdot \delta)(\sigma) = (w_H \cdot \delta)(j_*(\sigma)) = (w_H \cdot \delta)(1/|W_\tau|w_H \cdot \tau) = |W|/|W_\tau|\delta(w_H \cdot \tau) = |W|.$$

5.3.4. Imposing now some additional smoothness conditions and combining Theorem 5.4 with Theorem 3.2' we strengthen Theorem 5.4 as follows.

Theorem 5.5. *Under the conditions of Theorem 3.2' assume, in addition, that* $\dim M^H \leq \dim S^H + \dim W(H)$ *for any* $(H) \in \mathrm{Or}(M \setminus A)$. *Suppose that there exists an equivariant map* $f : A \to S$.

(a) *An equivariant extension of the map* f *over* M *exists.*

(b) *Let* $\Phi, \Psi : M \to S$ *be a pair of equivariant extensions of* f. *If* G *is infinite and* $(T) \notin O_{**}(M \setminus A, S)$ *then* $\deg \Phi = \deg \Psi$. *If* $(T) \in O_{**}(M \setminus A, S)$ *then there exist unique integer* α *depending only on the* G-*actions on* M *and* S , *and integers* $b(K) = b(K, \Phi, \Psi)$, $(K) \in O_{**}(M \setminus A, S, T)$, *such that*

$$\deg \Phi - \deg \Psi = \alpha \cdot \sum_{(K) \in O_{**}(M \setminus A, S, T)} \alpha(K, T) \cdot b(K) \cdot \chi(G/K).$$

If $(H) \in O_*(M \setminus A, S, T)$ then

$$\deg(\Phi|M^H) - \deg(\Psi|M^H) = \sum_{(K) \in O_{**}(M \setminus A, S, H)} \alpha(K, H) \cdot b(K) \cdot |N(H, K)|.$$

(c) Assume that A is a G-ENR. If $\{b(K)\}$, $(K) \in O_{**}(M \setminus A, S, T)$, is an arbitrary set of integers then for any equivariant extension $\Psi : M \to S$ of the map f there exists an equivariant extension $\Phi : M \to S$ of the map f such that for any $(H) \in O_{**}(M \setminus A, S, T)$

$$\deg(\Phi|M^H) - \deg(\Psi|M^H) = \sum_{(K) \in O_{**}(M \setminus A, S, H)} \alpha(K, H) \cdot b(K) \cdot |N(H, K)|$$

(d) Assume, in addition, that $\dim M^H < \dim S^H + \dim W(H)$ for all $(H) \notin O_{**}(M \setminus A, S, T)$ with $\dim W(H) > 0$. Let $\Phi, \Psi : M \to S$ be equivariant extensions of f and let $\{b(K)\}$ be a family of integers provided by (b). For all $(H) \notin O_{**}(M \setminus A, S, T)$ with $\dim W(H) = 0$ and $\dim M^H = \dim S^H$ assume that $\deg \Phi|M^H = \deg \Psi|M^H$. Then the statement

"Φ and Ψ are equivariantly homotopic if and only if $b(K) = 0$ for all $(K) \in O_{**}(M \setminus A, S, T)$"

is equivalent to condition (5.25) with M instead of X and S instead of Y.

Remark 5.7.   Using the methods of Chapter 4 one can easily formulate and prove infinite dimensional variants of Theorem 5.5(c,d) for compact vector fields commuting with one (linear, isometrical) action of a compact Lie group (see Theorem 4.2). We omit the obvious reformulation.

Suppose that under the conditions of Theorem 5.5 $A = \{x \in M \mid \dim M^{G_x} > \dim S^{G_x} + \dim W(G_x)\}$. By Corollary 1.3 (see also [Di1], [Ja1]) in such a case A is a G-ENR-space. Suppose, in addition, that $S^G \neq \emptyset$, and denote by $\mu : M \to S^G$ a trivial equivariant map such that $\mu(M) = \text{pt} \in S^G$. Set $f = \mu|A$.

Corollary 5.3.

(a) If G is infinite and $(T) \notin O_{**}(M, S)$ then for any equivariant map $\Phi : M \to S$ $\deg \Phi = 0$.

(b) Suppose that $(T) \in O_{**}(M \setminus A, S)$ and let $\Phi : M \to S$ be an equivariant extension of f. Then there exist a unique integer α depending only on the actions of G on M and S, and integers $b(K) = b(K, \Phi)$, $(K) \in O_{**}(M \setminus A, S, T)$, such that

$$\deg \Phi = \alpha \cdot \sum_{(K) \in O_{**}(M \setminus A, S, T)} \alpha(K, T) \cdot b(K) \cdot \chi(G/K).$$

If $(H) \in O_{**}(M \setminus A, S, T)$ then

$$\deg(\Phi|M^H) = \sum_{(K) \in O_{**}(M \setminus A, S, H)} \alpha(K, H) \cdot b(K) \cdot |N(H, K)|.$$

(c) Let $b(K)$, $(K) \in O_{**}(M \setminus A, S, T)$, be an arbitrary set of integers. Then there exists an equivariant extension $\Phi : M \to S$ of f such that for any $(H) \in O_{**}(M \setminus A, S, T)$

$$\deg(\Phi|M^H) = \sum_{(K) \in O_{**}(M \setminus A, S, H)} \alpha(K, H) \cdot b(K) \cdot |N(H, K)|.$$

(d) Assume, in addition, that $\dim M^H < \dim S^H + \dim W(H)$ for all $H \notin O_{**}(M \setminus A, S, T)$ with $\dim W(H) > 0$. Let $\Phi : M \to S$ be an equivariant extension of f and let $\{b(K)\}$ be a set of integers provided by (b). Assume that $\deg \Phi|M^H = \deg \Psi|M^H$ for all $(H) \notin O_{**}(M \setminus A, S, T)$ with $\dim W(H) = 0$. Then the statement

" Φ is equivariantly homotopic to μ if and only if $b(K, \Phi) = 0$ for all $(K) \in O_{**}(M \setminus A, S, T)$."

is equivalent to condition (5.25) with M instead of X and S instead of Y.

We complete this subsection with considering abelian group actions. Combining Corollary 3.10 with Corollary 5.3 yields the following result.

Corollary 5.4. *Suppose that under the conditions of Corollary 3.10 $S^G \neq \emptyset$. Let, further, $\dim M^H \leq \dim S^H + \dim W(H)$ for all $(H) \in \mathrm{Or}(M)$ and let $\mu : M \to S$ be as above.*

(a) For any equivariant map $\Phi : M \to S$ one can pick up integers $\alpha(H_j, H_{j+1})$, $j = 1, \ldots, s - 1$, and integers b_j, $j = 1, \ldots, s$, in such a way that for any H_i

$$\deg(\Phi|M^{H_i}) = \sum_{j<i} \alpha(H_j, H_{j+1}) \cdot \ldots \cdot \alpha(H_{i-1}, H_i) \cdot b_j \cdot |G/H_j| + b_i \cdot |G/H_i|.$$

(b) For any set of integer numbers b_j, $j = 1, 2, \ldots, s$, one can pick up integers $\alpha(H_j, H_{j+1})$, $j = 1, \ldots, s - 1$, and construct an equivariant map $\Phi : M \to S$ in such a way that for any H_i

$$\deg(\Phi|M^{H_i}) = \sum_{j<i} \alpha(H_j, H_{j+1}) \cdot \ldots \cdot \alpha(H_{i-1}, H_i) \cdot b_j \cdot |G/H_j| + b_i \cdot |G/H_i|,$$

where integers $\alpha(H_j, H_{j+1})$ are unique modulo $|H_j/H_{j+1}|$ and depend only on the G-actions on M and S.

(c) Assume that $\dim M^H < \dim S^H + \dim W(H)$ for all $(H) \notin O_{**}(M, S, T)$. Let $\Phi : M \to S$ be an equivariant map and let $\{b_j\}$ be a family of integers provided

by (a). Then Φ is equivariantly homotopic to μ if and only if $b_j = 0$ for all $j = 1, \ldots, s$.

Note that if G is an abelian group then the condition of type (5.25) is valid automatically (see Remark 5.6). Moreover, if H is a subgroup of non-maximal rank then $\dim W(H) > 0$ (cf. Corollary 5.3(d)).

5.3.5.   Proof of Theorem 5.4.

We order the orbit types of X in a standard way so that $(H_i) < (H_j)$ implies $i > j$ (see Chapter I).

(a) The first statement of Theorem 5.4 is a direct consequence from Theorem 1.3.

(b) This statement follows immediately from Basic Lemma (i) and Proposition 5.3.

(c) From the exactness of the cohomological sequence of the pair $(X^H, X^{>H})$ it follows that $j^* : H^{n(H)}(X^H, X^{>H}) \to H^{n(H)}(X^H)$ is an epimorphism. Hence there exists an $\alpha \in H^{n(H)}(X^H, X^{>H})$ such that $j^*(\alpha) = \beta$, and one gets

$$j^*(w_H \cdot \alpha) = w_H \cdot j^*(\alpha) = w_H \cdot \beta . \tag{5.26}$$

From Basic Lemma (ii) it follows that there exists a pair of $W(H)$-equivariant extensions $f_1^H, f_2^H : X^H \to Y^H$ of f such that $\omega^{n(H)}(f_1^H, f_2^H) = w_H \cdot \alpha$. Now from (5.26) and Proposition 5.3 we have:

$$(f_1^H)^*(O_H) - (f_2^H)^*(O_H) = j^*(\omega^{n(H)}(f_1^H, f_2^H)) = j^*(w_H \cdot \alpha) = w_H \cdot \beta .$$

(d) If $f_1, f_2 : X \to Y$ are equivariantly homotopic maps then

$$(f_1|X^H)^*(O_H) = (f_2|X^H)^*(O_H) \quad \text{for all } H \in \mathrm{Iso}(X) .$$

Conversely, let $(H) = (H_i)$ and let an equivariant map

$$f : \bigcup_{j<i} G(X^{H_j}) = X_{i-1} \to Y_{i-1} = \bigcup_{j<i} G(Y^{H_j}) \subset Y_i = \bigcup_{j\le i} G(Y^{H_j})$$

be defined. By Lemma 1.15 there exists a one-to-one correspondence between the relative G-equivariant homotopy classes of the extensions of f over

$$X_i = \bigcup_{j\le i} G(X^{H_j})$$

and the relative $W(H)$-equivariant homotopy classes of the extensions of $f|X^{>H}$ over X^H. Suppose that f_1^H, f_2^H are two $W(H)$-equivariant extensions of the map $f|X^{>H}$ over X^H. If $W(H)$ is infinite then

$$\dim\left(X_H/W(H)\right) = n(H) - \dim W(H)$$

and hence (assumptions on Y^H) the maps f_1^H and f_2^H are $W(H)$-equivariantly homotopic. Similarly, f_1^H and f_2^H are $W(H)$-equivariantly homotopic if $W(H)$ is finite and $\pi_{n(H)}(Y^H) = 0$. Thus we can assume below that $H \in \Phi(G, X, Y)$.

If $(f_1^H)^*(O_H) = (f_2^H)^*(O_H)$ then according to Basic Lemma (i) and Proposition 5.3 we have:

$$0 = (f_1^H)^*(O_H) - (f_2^H)^*(O_H) = j^* \omega^{n(H)}(f_1^H, f_2^H) = j^* w_H \cdot \alpha \ .$$

If condition (5.25) holds then $w_H \cdot \alpha = 0$ and from Basic Lemma (iii) and Corollary 5.2 it follows that there exists an equivariant homotopy (relative to $X^{>H}$) between f_1^H and f_2^H. On the other hand, if $w_H \cdot \alpha \neq 0$ then by the same lemma and corollary the extensions f_1^H and f_2^H are not homotopic.

The theorem is proved.

5.3.6.   Proof of Theorem 5.5. Statement (a) follows from Theorem 1.3. Statement (b) is a consequence of Theorem 3.2'.

To prove (c) note that since A is a G-ENR there exists a closed invariant neighborhood U of A such that any two equivariant extensions of f over U are equivariantly homotopic (cf. e. g. [Di1]). We can pick up the neighborhood U in such a way that the pair (M, U) is a relative G-CW complex. Assume by induction that for some $(H) \in O_{**}(M \backslash A)$ there exists a $W(H)$-equivariant extension $\Phi_H : M^{>H} \bigcup U^H \to S$ of $f|U^H$ such that

$$\deg(\Phi_H|M^L) - \deg(\Psi|M^L) = \sum_{(K) \geq (L),\ (K),(L) \subset O_{**}(M \backslash A)} \alpha(K, L) \cdot b(K) \cdot |N(K, L)|$$

for any $(L) > (H)$. Let $\Phi^H : M^H \to S^H$ be an arbitrary $W(H)$-equivariant extension of Φ_H. From the proof of Theorems 3.1 and 3.2 it follows

$$\deg \Phi^H - \deg(\Psi|M^H) = \sum_{(K) > (H),\ (K) \in O_{**}(M \backslash A)} \alpha(K, H) \cdot b(K) \cdot |N(H, K)| + p(H),$$

where $p(H) = c(H) \cdot |W(H)|$ and $c(H)$ is an integer. If $c(H) \neq b(H)$ then using Theorem 5.4(c) we can find an equivariant extension Φ' of Φ_H such that

$$\deg \Phi' - \deg \Phi^H = (b(H) - c(H)) \cdot |W(H)|$$

from which (c) follows immediately.

Finally, (d) is a direct consequence of Theorem 5.4.

5.3.7.   To illustrate the methods developed in the present and in the third chapters we consider certain examples.

Example 5.2.   Let $q > 2$ be an integer and let $G = Z_2 \times Z_q =< a > \times < b >$. Let P be a two-dimensional representation of the group G given by the rule $a \rightarrow e^{\pi i}, b \rightarrow 1$ and let Q be a two-dimensional representation of the group G given by the rule $a \rightarrow 1, b \rightarrow e^{2\pi i/q}$. Set $V = P^m \bigoplus Q$, $m > 0$. This representation induces an orthogonal action of G on an $(2m + 1)$-dimensional sphere $S(V)$. Take also the action of Z_2 on S^1 given by the reflection with the fixed point set $S^0 = \{x, y\}$, and the action of Z_q on a torus T^{2m} which rotates each of the $2m$ circles of the cartesian product by the angles $2\pi ki/q$, $k = 0, 1, ...q - 1$. These two actions induce an action of G on $T^{2m+1} = T^{2m} \times S^1$ such that $F_1 = (T^{2m+1})^{Z_2} = T^{2m} \times \{x\} \bigcup T^{2m} \times \{y\}$. The group Z_q acts on F_1 and on $F_2 = S(V)^{Z_2} = S^1$ "in the same way", hence the projection $p : T^{2m} \rightarrow S^1$ is Z_q-equivariant. Therefore we have a Z_q-equivariant map $f_0 : F_1 \rightarrow F_2$. It is easy to see (cf. e.g Theorem 1.3) that f_0 can be extended to a G-equivariant map $f : T^{2m+1} \rightarrow S(V)$, and since $\dim F_1 > \dim F_2$ we have by Theorem 3.1 that for any G-equivariant maps $\Phi, \Psi : T^{2m+1} \rightarrow S(V)$ the following congruence is true: $\deg \Phi - \deg \Psi \equiv 0$ (mod $2q$). In other words, for any G-equivariant map $\Phi : T^{2m+1} \rightarrow S(V)$ we have $\deg \Phi = d + 2qk$, $k \in Z$, where the integer d is independent of Φ. We also know from Theorem 5.5 that for any $k \in Z$ there exists an equivariant map $\Psi : T^{m+1} \rightarrow S(V)$ such that $\deg \Psi = d + 2qk$. To find the number d let us construct an equivariant map $f_1 : T^{2m} \times S^1 \rightarrow T^{2m} \times S^1$ of degree 0. Indeed, let $I : T^{2m} \rightarrow T^{2m}$ be the identity map and let $\theta : S^1 \rightarrow S^1$ maps S^1 into the point $x \in (S^1)^{Z_2}$. Clearly, the map $f_1 = I \times \theta$ is equivariant and $\deg f_1 = 0$. Finally, we have $\deg(f \circ f_1) = 0$ and hence $d = 0$.

Remark 5.8.   We can consider the same situation for the action of the group $Z_2 \times T^k$, $k > 0$. As it follows from Theorem 3.1 the degree of any equivariant map in this case must be equal to zero.

Example 5.3.   Take a faithful representation of a finite group H in a k-dimensional vector space W where the number $k > 0$ is even. Consider a representation of a group $G = Z_2 \times H =< a > \times H$ in a two-dimensional vector space P given by the rule $a \rightarrow e^{\pi i}$, $H \rightarrow 1$. As in Example 5.2 we have a representation of G in W with the kernel Z_2. Let m be another positive even number and let $t = (1/2) \cdot [(m - 1) \cdot (k - 1) + 1]$. Set $V = P^t \bigoplus W$. This representation induces an orthogonal action of G on a $(2 \cdot t + k - 1)$-dimensional sphere $S(V)$. Take also an action of Z_2 on S^1 given by a reflection with a fixed point set $S^0 = \{x, y\}$ and a "coordinatewise" action of the group H in $S(W)^m$. These two actions induce an action of G on $Q = S(W)^m \times S^1$ such that $F_1 = Q^{Z_2} = S(W)^m \times \{x\} \bigcup S(W)^m \times \{y\}$. The group $H = W(Z_2)$ acts on F_1 and on $F_2 = S(V)^{Z_2} = S(W)$ and we have an H-equivariant map $f_0 : F_1 \to F_2$ induced by the projection of $S(W)^m$ onto $S(W)$. Let us verify that this map has an equivariant extension over Q. Indeed, in accordance with the general extension scheme described in Lemma 1.15 we need to look at subgroups K in G such that K does not contain Z_2 and therefore, $K \subset H$. Suppose that $\dim S(W)^K = r$, i. e $\dim W^K = r + 1$. We have $\dim Q^K = r \cdot m + 1$ and $\dim S(V)^K = \dim V^K - 1 = r + 2 \cdot t$. Hence $\dim Q^K$ is strictly less than $\dim S(V)^K$ if $r < k - 1$ and $\dim Q^K = \dim S(V)^K$ if $r = k - 1$. Since the action of H on $S(W)$ is effective, the latter can happen only if $K = \{1\}$. Applying Theorems 3.1 and 5.5 in the same way as in Example 5.2 we find that:

(a) any equivariant extension of f_0 over Q has a degree equal to $2 \cdot l \cdot |H|$ for some integer l;

(b) for any integer l there exists an extension of f_0 of degree $2 \cdot l \cdot |H|$.

Example 5.4 (cf. [IV1]). Let V be a four-dimensional complex vector space with coordinates X_1, X_2, X_3, X_4. Let the group $G = Z_{12}$ act on V in a standard way with kernel $H = Z_3$, i. e. $S(V)$ is a free Z_4-representation sphere. Denote by V_1 a two-dimensional complex vector space and assume that Z_{12} acts in a standard way on V_1 with kernel $K = Z_2$. Let, further, W be a three-dimensional complex vector space with coordinates U_1, U_2, U_3 where Z_{12} acts in a standard way with kernel $Z_6 = HK$. Finally, denote by W_1 another three-dimensional complex where Z_{12} acts in a standard way with kernel $\{e\}$.

The above actions induce Z_{12}-actions on $A = V \oplus V_1$ and $B = W \oplus W_1$ in a natural way. We are interested in the degrees of Z_{12}-equivariant maps $S(A) \to S(B)$. Define a map $f : V \to W$ by

$$U_1 = X_1^2 - (X_2^2)^*;$$

$$U_2 = X_3^2 - (X_4^2)^*;$$

$$U_3 = \text{Re}\,(X_1 \cdot X_2) + i \cdot \text{Re}\,(X_3 \cdot X_4)$$

(here the symbol "*" denotes the complex conjugation). It is easy to see that f is G/H-equivariant. Moreover, f has no zeros on $S(V)$ so that f gives rise to a map $S(V) \to S(W)$ which we will denote by the same letter f.

Since $\dim V_1 < \dim W$ there exists (by Theorem 1.3) a G/K-equivariant map $g : S(V_1) \to S(W)$. Since (H) and (K) are incomparable the pair (f, g) gives rise to a Z_{12}-equivariant map $\varphi : S(V) \bigcup S(V_1) \to S(W)$. Using Theorem 1.3 once again we can extend φ to an equivariant map $\Phi_0 : S(A) \to S(B)$. Since $S(A^H) = S(V)$,

$S(A^K) = S(V_1)$, $S(B^H) = S(B^K) = S(W)$ we have $\dim S(A^H) > \dim S(B^H)$
and $\dim S(A^K) < \dim S(B^K)$. Hence (Corollary 3.2), $\deg \Phi_0 \equiv 0 \pmod 2$ and
$\deg \Phi_0 \equiv 0 \pmod 3$, i. e. $\deg \Phi_0 \equiv 0 \pmod 6$. It turns out that there exists an
equivariant map $\Phi : S(A) \to S(B)$ of degree exactly 6 (see [IV1] for details).

Observe that $O_{**}(S(A), S(B)) = \{e\}$. Hence, from Theorem 3.1 it follows
immediately that for any equivariant map $\Psi : S(A) \to S(B)$ one has $\deg \Psi \equiv 6$
(mod 12).

We can also take an arbitrary "reasonable" manifold M with the trivial Z_{12}-
action and consider G-manifolds $P = M \times S(A)$ and $Q = M \times S(B)$. By Corollary
1.6 we have $Q_{**}(P, Q) = \{(e), (G)\}$. Therefore (Theorem 3.1), for any equivariant
map $\lambda : P \to Q$ we have:

$$\deg \lambda = \alpha(G, e) \cdot \deg(\lambda | M) = 6 \cdot \deg(\lambda | M) \pmod{12}$$

since we can let $\alpha(G, e) = \deg \Phi$, where Φ is from the above example.

We complete this example with the following simple observation. Let $H \subset Z_4$
be the subgroup of order two. Let us consider the "factor-action" of the group
$Z_2 = Z_4/H$ on the real projective space $RP^7 = S^7/H$ (in homogeneous coor-
dinates $\{\text{Re } X_i, \text{ Im } X_i, i = 1, 2, 3, 4\}$ this action is given by $(\text{Re } X_i, \text{ Im } X_i) \to$
$(-\text{Im } X_i, \text{ Re } X_i)$. Since H acts trivially on W, any equivariant map $S(V) \to S(W)$
is in fact a composition of the natural projection (the Hopf map) $S^7 \to S^7/H = RP^7$
with some Z_4/H-equivariant map $RP^7 \to S(W)$. In particular, the map f described
above provides an example of a Z_2-equivariant map $RP^7 \to S(W)$.

We are grateful to J. Ize for giving us a permission to reproduce (a part of) his
example (see [IV1]) in our notes.

Example 5.5 (cf. [Pe]). Let p, q be two different odd prime numbers. Let $G = PQ$
be a semidirect product of a cyclic group of order p by a cyclic group of order q. Assume that $N(Q) = Q$ (as a generic example of such a situation, consider a subgroup
Q of prime order q of the group of automorphisms Aut P). Linear representations
of this group are easily classified (cf. e.g. [Se]). For our purposes we will consider
only complex representations of G of the following structure. A typical irreducible
G-representation U is induced by a faithful one-dimensional representation of P
and therefore has the dimension q (see, for instance, [Se]). It is easy to see that
$\dim_C U^Q = 1$ and that P acts freely on $S(U)$ (note that as a P-module U splits
into a direct sum of q one-dimensional representations which are permuted by Q).
Hence, the G-action on $S(U)$ has only two orbit types (Q) and (e). Therefore, taking
a direct sum of s such representations (denote this sum by V) we get a G-action on

the sphere $S(V)$ $(\dim(S(V)) = t = 2sq - 1)$ such that the only non-free orbit type is (Q) and $\dim(S(V)^Q) = 2s - 1$. From this description of the G-representations it follows easily that for any G-representation W having the same orbit structure and the same dimension as V, we have $\dim(S(W)^Q) = 2s - 1$.

Proposition 5.4.

(a) *Any continuous map* $f : S(V)^Q \to S(W)^Q$ *can be extended equivariantly over* $S(V)$;

(b) *for any equivariant map* $g : S(V) \to S(W)$ *one has* $\deg f \not\equiv 0 \pmod{p}$;

(c) *there exists an equivariant map* $h : S(V) \to S(W)$ *such that* $\deg h \equiv 0 \pmod{q}$.

Proof. Since $|W(Q)| = 1$ statement (a) follows from Theorem 1.3. Statement (b) is a trivial case of Corollary 2.7. Statement (c) trivially follows from statement (a) and Corollary 3.2.

To make this example more interesting let us recall the following result of T. Petrie [Pe].

Theorem P. *For any* $l = 2kq - 1$ $(k \in N)$ *there exists a smooth free G-action on the sphere* S^l.

Take positive integers $a > 1$ and $b > 1$ and set $2q(a + b) - 1 = t$. Take the G-representation A with $\dim_C A = qa$ of the type described above (i. e. P acts freely on $S(A)$ and $\dim(S(A)^Q) = 2a - 1$. Take also the G-representation V of "the same" type with $\dim_C V = q(a + b)$. Further, take the free G-action on S^{2bq-1} provided by Theorem P and consider the join $S(A) * S^{2bq-1} = S^t$ with the natural G-action on it. Clearly, this G-action is not smooth. However, $(S^t)^Q$ is a sphere of dimension $2b - 1$ and the P-action on S^t is free.

Proposition 5.5.

(a) *Any continuous map* $f : S(V)^Q \to (S^t)^Q$ *can be equivariantly extended over* $S(V)$;

(b) *for any equivariant map* $f : S(V) \to S^t$ *the following relations hold:*

(b$_1$) $\deg f \not\equiv 0 \pmod{p}$;

(b$_2$) $\deg f \equiv 0 \pmod{q}$.

Proof. The only statement which is really new here is (b$_2$). Let W be a Q-representation such that $\dim S(W) = t$, the Q-action on $S(W)$ is semifree and dim $S(W)^Q = \dim(S^t)^Q$. Take a homeomorphism $h : (S^t)^Q \to S(W)^Q$ and extend it over S^t to get a Q-equivariant map $f_0 : (S^t) \to S(W)$. In the same way, let $f_1 : S(W) \to S^t$ be a Q-equivariant extension of the homeomorphism h^{-1}. It is clear (see, for instance, Corollary 3.2) that $\deg(f_0 \circ f_1) \equiv 1 \pmod{q}$. Hence $\deg f_0 \not\equiv 0 \pmod{q}$. Using Corollary 3.2 once again we get $\deg(f_0 \circ f) \equiv 0 \pmod{q}$ which proves (ii).

5.4. Existence of equivariant maps

5.4.1.   K. Borsuk in [Bor2], [Bor3] (see also [LS]) proved that if S^k and S^ℓ are k- and ℓ-dimensional spheres respectively and there exists an odd map from S^k into S^ℓ then $k \leq \ell$.

This result was generalized by A. Dold [Do2] to a situation of free actions of an arbitrary finite group on spheres.

Later on J. Daccach obtained A. Dold's Theorem for free G-manifolds. To formulate J. Daccach's result we need the following definition.

Definition (see [Dac]). Suppose that a finite non-trivial group G acts freely on an m-dimensional manifold M and there exist a free action of G on the m-dimensional sphere S^m and an equivariant map $f : M \to S$ such that $\deg f$ is relatively prime to $|G|$. Then M is called a manifold with property (Γ).

Proposition 5.6 (see [Dac]). *Let M be an m-dimensional manifold with property (Γ) and N be an n-dimensional free G-manifold. If there exists an equivariant map $f : M \to N$ then $m \leq n$.*

We generalize Daccach's result to a situation of arbitrary (non-free) actions; in addition, N may be an arbitrary metric space. It is especially important to note (in contrast to [Dac]) that we do not use the construction of the trace of an action introduced by D. Gottlieb [Got].

5.4.2.   We need the following definition.

Definition.   *Let a non-trivial finite group G act on a topological (compact, connected, oriented) n-dimensional manifold M with orbit types $(H_1), \ldots, (H_\ell)$. We say that M satisfies property (Γ') if there exists an action of G on the n-dimensional sphere S such that:*

1) the set S^{H_i} is locally and globally k-connected for every $k = 0, 1, \ldots, n_i$, where $i = 1, \ldots, \ell$ and $\dim M_{(H_i)} \le n_i + 1$;

2) there exists an equivariant map $\Psi : M \to S$ for which $\deg \psi$ is relatively prime to $\mathrm{GCD}\ \{|G/H_i|\}_{i=1}^{\ell} = \alpha(G)$.

Theorem 5.6.   *Let M be a manifold with property (Γ') and $\alpha(G) > 1$. Suppose that N is a metric G-space, $(D_1), \ldots, (D_r)$ are orbit types in N, and for every $s = 0, 1, \ldots, n_j$, $j = 1, \ldots, r$, the sets S^{D_j} are locally and globally s-connected and $\dim N_{(D_j)} \le n_j + 1$. If there exists an equivariant map $f : M \to N$ then $H_n(N) \ne 0$ (in particular, $n \le \dim N$).*

Proof According to the assumptions on N and S and Theorem 1.3 there exists an equivariant map $\Psi : N \to S$. If $H_n(N) = 0$ then $\deg\ (\psi \circ f) = 0$ but this contradicts the assumptions on M and Theorem 2.1.

Remark 5.9.   Assume M is a smooth manifold and G is an arbitrary compact Lie group acting smoothly on M. Then one can easily reformulate Theorem 5.6 in a form compatible with Theorems 3.1 and 3.2. We omit the obvious reformulations.

5.5. Atiyah-Tall Theorem and related topics

5.5.1.   In this section we will discuss the well-known theorem of Atiyah-Tall (see [AT]). We will start with a few necessary definitions from representation theory of finite groups (cf. e. g. [Se]).

Recall that if a group G is of order l, then any representation of G over a field of characteristic 0 can be realized over the field $Q(\omega)$ where Q is the field of rationals and ω is a primitive l-th root of unity. In particular, the values of a character of a a complex or real representation of G belong to $Q(\omega)$.

Definition.   *Two complex or real representations of a finite group G ($|G| = l$) are said to be conjugate if their characters are conjugate by an element of the Galois group $\Gamma_l = Gal(Q(\omega)/Q)$.*

The following result is well known.

Theorem 5.7R (see [AT]). *Let V and W be real orthogonal representations of a finite p-group G, $|G| = l$, $\dim V = \dim W$, $p > 2$. Let $S(V)$ and $S(W)$ be the unit spheres in V and W. There exists an equivariant map $f : S(V) \to S(W)$ such that $\deg f \not\equiv 0$ (mod p) if and only if irreducible components of V and W are conjugate in pairs by elements (possibly different) of the group Γ_l.*

The complex analog of this theorem can be stated as follows.

Theorem 5.7C (see [AT, Sn]). *Let G be a finite p-group, $|G| = l$, and let V and W be complex unitary G-representations such that $\dim V = \dim W$. Let $S(V)$ and $S(W)$ be the unit spheres in V and W. There exists an equivariant map $f : S(V) \to S(W)$ such that $\deg f \not\equiv 0$ (mod p) if and only if irreducible components of V and W are conjugate in pairs by elements (possibly different) of the group Γ_l.*

Remark 5.10. In their paper [AT], M. Atiyah and D. Tall have proved Theorems 5.7R and 5.7C under the assumption that $p > 2$. Subsequently, V. Snaith extended the result to include the case $p = 2$ for complex representations [Sn]. Various generalizations of this result to the case of arbitrary compact Lie groups can be found in [LW].

Observe that the original proof by M. Atiyah and D. Tall has been based on the use of λ-rings and Adams operations. We will give a simple proof of Theorems 5.7R and 5.7C based on the results of the previous chapters, and discuss some (apparently new) generalizations which deal with equivariant maps from a manifold into a sphere.

We will consider Theorem 5.7C first, so let us assume all representations to be complex unless stated otherwise.

5.5.2.   We will start the proof of Theorem 5.7C with the following

Lemma 5.3 (cf. [LW]). *An equivariant map $f : S(V) \to S(W)$ with $\deg f \not\equiv 0$ (mod p) exists if and only if $\dim V^H = \dim W^H$ for any subgroup H of G.*

Proof of Lemma 5.3. Let $f : S(V) \to S(W)$ be an equivariant map. If for some H $\dim V^H \neq \dim W^H$ then $\deg f \equiv 0$ (mod p) by Corollary 3.2. Assume therefore that $\dim V^H = \dim W^H$ for all subgroups H. In this case we can proceed for example as follows. By Theorem 1.3 an equivariant map $f : S(V) \to S(W)$ exists and if $V^G = \emptyset$ then by Corollary 2.8 $\deg f \not\equiv 0 \pmod{p}$. Finally, if $V^G \neq \emptyset$ then let $f_0 : V^G \to W^G$ be a map of degree one and let $f : S(V) \to S(W)$ be an equivariant extension of f_0 (it exists by Theorem 1.3). Again by Corollary 3.2 we have $\deg f \equiv \alpha$ $\pmod{p}$ where α is the degree of an equivariant map $S((V^G)^\perp) \to S((W^G)^\perp)$ and "$\perp$" stands for orthogonal complements. It remains to observe that by the above argument $\alpha \not\equiv 0 \pmod{p}$.

Remark 5.11.   An analysis of the proof of Lemma 5.3 shows that if G is a p-group and X and Y are smooth G-manifolds (with some usual conditions (see Subsection 3.1.5)) then for the existence of an equivariant map $f : X \to Y$ with $\deg f \not\equiv 0$ $\pmod{p}$ it is necessary that $\dim X^H = \dim Y^H$ for all H.

Remark 5.12.   It is also easy to see that under the conditions of Lemma 5.3 the orbit structures of $S(V)$ and $S(W)$ are the same, i. e. if H is a stationary subgroup for $S(V)$ then it is also a stationary subgroup for W.

Lemma 5.4.   *If representations V and W are direct sums of pairwise conjugate irreducible representations then $\dim V^H = \dim W^H$ for any subgroup H.*

   Proof of Lemma 5.4. The statement is obvious for a pair of conjugate representations. Since the functor $(\cdot)^H$ commutes with direct sums $((A \oplus B)^H = A^H \oplus B^H)$ the lemma follows by induction on the number of conjugate pairs.

   From now on we assume that there exists an equivariant map $f : S(V) \to S(W)$ such that $\deg f \not\equiv 0 \pmod{p}$ and we fix this map f. By Lemma 5.3 $\dim V^H = \dim W^H$ for any subgroup H of G. We will use the phrase "V and W are as required" instead of "V and W are sums of pairwise conjugate irreducible representations".

Lemma 5.5.   *Under the assumptions of Theorem 5.7C one can assume without loss of generality that for any proper normal subgroup N of G the set $S(V^N)$ is empty.*

   Proof of Lemma 5.5. Take a proper normal subgroup N of G such that $S(V^N) \neq \emptyset$ and set $B = (V^N)^\perp$, $D = (W^N)^\perp$. By Corollary 3.2, $\deg f \equiv \alpha \cdot (\deg f|S(V^N)) \pmod{p}$ where α is the degree of some N-equivariant map $S(B) \to S(D)$. Hence $\deg f|S(V)^N \not\equiv 0 \pmod{p}$. Since V^N and W^N are G-invariant and $f|S(V)^N$ is G-equivariant we get that $\dim(V^N)^H = \dim(W^N)^H$ for any subgroup H of G (see Lemma 5.3). It is clear also that for any subgroup H of G we must have $\dim B^H = \dim D^H$ as well. Applying the same arguments to G-representation pairs (V^N, W^N) and (B, D) we get

$$V = \oplus_i V_i, \qquad W = \oplus_i W_i,$$

where the G-subresentations V_i and W_i satisfy the following conditions:

a) $\dim V_i = \dim W_i$;

b) there exists an equivariant map $f_i : S(V_i) \to S(W_i)$ with $\deg f_i \not\equiv 0$ (mod p);

c) for any proper normal subgroup N of G one has $S(V_i^N) = \emptyset$.

The lemma is proved.

From now on assume that for any proper normal subgroup N of G the set $S(V^N)$ is empty.

Lemma 5.6.   *If G is abelian then V and W are as required.*

Proof of Lemma 5.6. If G is abelian then by Lemma 5.5 G acts freely on $S(V)$ (and on $S(W)$). Hence G is cyclic. However, any two faithful irreducible representations of a cyclic group are conjugate. This completes the proof of Lemma 5.6.

5.5.3.   In addition to the above assumptions we can now assume that G is non-abelian, and therefore V (and, hence W) is a sum of faithful irreducible representations of dimension greater than one.

Let V_0 be an irreducible summand of minimal dimension – we take all the irreducible summands in V and W and take a smallest one. Assume that $V_0 \subset V$.

It is well known (see, for instance, [Se]) that V_0 is induced by a one-dimensional representation of a subgroup $H < G$. Let $\lambda : H \to S^1$ be the corresponding homomorphism, and let $v \in V_0$ be a unit vector (i.e. $v \in S(V_0) \subset S(V)$) such that $hv = \lambda(h)v$ for any $h \in H$. Note that $\dim V_0 = |G/H|$. Let H_0 be the kernel of the homomorphism λ, i. e. $\lambda(h)v = v$ for all $h \in H_0$. Let us check that H_0 is a stationary subgroup of v. Indeed, if $gv = v$ for some $g \in G$ then, first of all, $g \in H$ (otherwise v and gv must be linearly independent since V_0 is induced by the representation of H in $C \cdot v$). Further, the representation of H/H_0 in $C \cdot v$ is faithful and therefore g must belong to H_0. It follows now from Remark 5.12 that there exists a vector $w \in S(W)$ such that H_0 is the stationary subgroup of w. Therefore, $S(W^{H_0})$ is non-empty and there exists $w \in S(W^{H_0})$ such that $G_w = H_0$. Since H

normalizes H_0, H acts on W^{H_0}, and we have the action of the cyclic group H/H_0 on W^{H_0}. Hence we can find a one-dimensional representation θ of H in W^{H_0} which we denote by $C \cdot w$. Again we have that $\theta(H)w = C \cdot w$, $\theta(H_0)w = w$ and H_0 is the stationary subgroup of w. Let W_0 be a representation of G spanned by w, i. e. $W_0 = C \cdot G(w)$. It is easy to see that

$$W_0 = \sum_{i=1}^{|G/H|} C \cdot (g_i w),$$

where g_i, $i = 1, 2, .., |G/H|$, is a representative coset of H in G. This is not necessarily a direct sum of one-dimensional subspaces, and if it is not a direct sum we have found an invariant subspace in W of dimension less than $|G/H| = \dim V_0$ – a contradiction. If, however, it is a direct sum then W_0 is induced by θ and therefore V_0 and W_0 is a pair of conjugate representations (their characters are conjugate by the automorphism which conjugates θ and λ).

What is left is to note that by Lemma 5.4 $\dim(V_0)^K = \dim(W_0)^K$ for any subgroup K in G, and if V_1 (respectively, W_1) is an orthogonal complement to V_0 (respectively, to W_0) in V (respectively, in W) then representations V_1 and W_1 satisfy the same conditions. Hence by Lemma 5.3 there exists an equivariant map $g : S(V_1) \to S(W_1)$ with $\deg g \not\equiv 0 \pmod{p}$. Hence, Theorem 5.7C is established by induction.

Remark 5.13.   In fact, we have proved that for a p-group G the following conditions are equivalent:

a) complex representations V and W are sums of pairwise conjugate pairs of irreducible representations;

b) $\dim V^H = \dim W^H$ for any subgroup H;

c) there exists a map $f : S(V) \to S(W)$ such that $\deg f \not\equiv 0 \pmod{p}$.

5.5.4.   Let us return to Theorem 5.7R. It is quite obvious that everything stated above for complex representations up to Lemma 5.6 is valid for real representations as well. Hence, we can begin the proof of Theorem 5.7R assuming that G is non-abelian and that V and W are sums of faithful irreducible representations of dimension greater than one. Since $p > 2$ any non-trivial irreducible representation of G is obtained by restricting the scalars of an irreducible complex representation. In this case a character of a real representation is equal to $c + c^*$ where c is a character of the corresponding complex representation and "*" means the complex conjugation (cf. e. g. [Se]). Hence we deduce in the same way as above that V and W are sums of pairwise conjugate complex representations. Let us take a pair of conjugate irreducible complex characters a and b, that is $b = a^g$ where $g \in \Gamma_l$. Since the complex conjugation belongs to Γ_l, we have

$$(b + b^*) = a^g + (a^g)^* = a^g + (a^*)^g = (a + a^*)^g,$$

and this completes the proof of Theorem 5.7R.

5.5.5.   We complete this section with generalizations of Theorem 5.7R dealing with equivariant maps from a manifold into a sphere.

Theorem 5.8.   Let G be a p-group $(p > 2)$ acting on a (compact, connected, oriented) n-dimensional smooth manifold M. Assume for simplicity that M^H is connected for all subgroups H of G. Let V be an orthogonal $(n + 1)$-dimensional G-representation. The following conditions are equivalent:

a) there exists an equivariant map $f : M \to S(V)$ such that $\deg f \not\equiv 0$ (mod p);

b) for any subgroup H in G with $\dim M^H = \dim S(V^H)$ there exists a $W(H)$-equivariant map $f_H : M^H \to S(V^H)$ such that $\deg f \not\equiv 0$ (mod p) and, moreover, for any point x with $G_x = H$ the H-representations in $N_x(M^H)$ and in $\tau_H = (V^H)^\perp$ split into sums of pairwise conjugate irreducible representations (here x is an arbitrary point with $G_x = H$ and $N_x(M^H)$ denotes the normal plane at x to M^H in M);

c) for any minimal orbit type $(H) \in \mathrm{Or}(M)$ (that is M^H does not contain any other M^K, $(K) \neq (H)$, $(K) \in \mathrm{Or}(M)$) with $\dim M^H = \dim S(V^H)$, there exists a $W(H)$-equivariant map $f_H : M^H \to S(V^H)$ such that $\deg f_H \not\equiv 0$ (mod p) and the H-representations in $N_x(M^H)$ and in $(V^H)^\perp$ split into sums of pairwise conjugate irreducible representations.

Proof.   a) $\Rightarrow$ b). This implication follows from Corollary 3.2 as described above.

The implication b) $\Rightarrow$ c) is obvious.

Let us show c) $\Rightarrow$ a). Take a minimal orbit type H. If $H = e$ then since we have a map f_H there is nothing to prove. Otherwise, take a small sphere σ_H orthogonal to M^H. By Lemma 5.4 for any subgroup K of H, $\dim(\sigma_H)^K = \dim S(((V^H)^\perp)^K)$, hence $\dim M^K = \dim S(V^K)$ and the same is true for all other orbit types. By our assumptions and by Theorem 5.7R there exists an equivariant map $f_0 : \sigma_H \to S((V^H)^\perp)$ such that $\deg f_0 \not\equiv 0$ (mod p). Set $\alpha_H = \deg f_0$. As before, there exists an extension f of the map f_H over M (Theorem 1.3). By Corollary 3.2 $\deg f \equiv \deg f_H \cdot \alpha_H$ (mod p).

Corollary 5.5. *Under the assumptions of Theorem 5.8 assume, in addition, that M^G is non-empty. Then the following statements are equivalent:*

a) there exists an equivariant map $f : M \to S(V)$ such that $\deg f \not\equiv 0 \pmod{p}$

b) $\dim M^G = \dim S(V^G)$ and the G-representations in $N_x(M^G)$, $x \in M^G$, and in $(V^G)^\perp$ are sums of pairwise conjugate orthogonal representations.

5.6. Historical and bibliographical notes

5.6.1. If $q = 1$ and φ is an odd function in u Theorem 5.1 was established by P. Rabinowitz [Ra1] (see also [Ni1]). The case when an arbitrary compact Lie group G acts in such a way that $(R^q)^G = \{0\}$, was considered by W. Marzantowicz [Mar1]). Theorem 5.1 and the corresponding example have been suggested by Z. Balanov and V. Ajevsky [BA1,BA2]. The scheme we follow in the proof of Theorem 5.1 has been suggested by P. Rabinowitz [Ra1] (see also [Ni1]).

5.6.2. Historically, the first category result is the famous Lusternik-Schnirelman Theorem claiming that the category of the n-dimensional real projective space equals $n + 1$ (see [LS]). In terms of genus the Lusternik-Schnirelman Theorem can be formulated as follows: the genus of the n-dimensional sphere with respect to the antipodal action is equal to $n + 1$. This result was generalized by A. Fet [Fe] to the case of an arbitrary free involution on the sphere. The situation of a free action of an arbitrary finite cyclic group was considered by M. Krasnoselskii [Kr3, KZ] in the framework of the geometric approach. For free actions of an arbitrary compact Lie group Theorem 5.2 was proved by T. Bartsch [Bar1] (see also [KB, Ba1, Ba2, Ba4] where a situation of a free action of a finite group was considered). For a free action of a finite group on a cohomological sphere Theorem 5.2 was proved by W. Marzantowicz [Mar2].

A. Švartz [Šv] was the first to consider the situation of a non-free action of a cyclic group on the sphere and obtained the particular case of Theorem 5.2 in this situation. It should be noticed that specific properties of cohomologies of the cyclic group with coefficients in a field were essentially used in [Šv]. For linear semi-free S^1-action Theorem 5.1 was proved by E. Fadell [F1] in the framework of the homological approach.

Theorems 5.2 and 5.3 were proved by Z. Balanov and S. Brodsky [BB3]. Some useful G-category estimates for (non-free) p-group actions on a sphere can be found in [IzM, Bar1, Mar2]. After Theorem 5.2 was obtained, T. Bartsch informed us that he has a homological proof of this result (using a reduction to Švartz's result mentioned above).

It should be noticed that the systematical study of the genus (G-category, G-index, cup-length etc.) of G-spaces in a non-free case has been started by E. Fadell, S. Husseini and P. Rabinowitz [Fa1-Fa3, FH1, FH2, FR] and V. Benci [Be]. Today the literature on this subject is incredibly rich, and we are not able to provide a sufficiently complete list of publications. To keep our monograph in a reasonable size we refer the reader for the additional bibliographical information to four recent monographs by P. Rabinowitz [Ra2], M. Struwe [Str], J. Mawhin and M. Willem [MW] and T. Bartsch [Bar1] (see also H. Steinlein's survey [St]).

5.6.3.   In the case of a simplicial action of a cyclic group of a prime order a version of the Basic Lemma was proved by S. Eilenberg about fifty years ago (see [Ei]).

Condition (5.25) obviously holds when $1 + \dim(X^{>H}) < n(H)$ for all $H \in \Phi(G, X, Y)$ (cf. [Di1], Lemma 8.4.1 and [Di2], Theorem 4.11, p. 126).

Suppose that $H^{n(H)}(X^H) = Z$ for all $H \in \Phi(G, X, Y)$ and denote by 1_H the corresponding generators. Then the homomorphism $e_{W(H),X} : W(H) \to \{+1, -1\}$ is correctly defined. If $e_{W(H),X} = e_{W(H),Y}$ then $w_H = |W(H)| \cdot O_{X^H}$ and we get the Equivariant Hopf Theorem in the form presented by T. tom Dieck in [Di2], p. 126 (cf. also [Di1, To, Lai]).

The case when X is a smooth n-dimensional manifold, $Y = R^{n+1} \setminus \{0\}$, G is a finite group acting smoothly on X and Y and $Y^H = Y$ for all $H \in \mathrm{Iso}(X)$ was considered by C. Bowszyc in [Bow1, Bow2].

Possible values for degrees of equivariant maps between unit spheres $S(V)$ and $S(W)$ in two linear representations of a finite group G were studied by J. Tornehave [To]. In the case when $\dim S(V)^H = \dim S(W)^H$ for any subgroup $H \subset G$, Tornehave gave a complete set of congruences satisfied by the numbers $\deg f|S(V)^H$ for an equivariant map $f : S(V) \to S(W)$. The condition $\dim S(V)^H = \dim S(W)^H$ is related to the connectedness condition of Theorem 5.4 and is necessary if one wants to apply equivariant cohomologies (cf. [Di1, Di2, To, Lai]). The case of linear actions of abelian groups has been studied intensively in [IV]. Theorem 5.4 was proved by Z. Balanov and A. Kushkuley [BK3].

Theorem 5.5 as well as Corollaries 5.3 and 5.4 were proved by A. Kushkuley who also suggested Examples 5.2, 5.3 and 5.5. Our considerations in Example 5.4 are based on an unpublished example by J. Ize (see [IV1]).

5.6.4.   Theorem 5.6 was suggested by Z. Balanov.

Certain sufficient conditions for the existence of a G-equivariant map from a cohomological sphere to a G-space (G is an arbitrary compact Lie group) have been suggested by W. Marzantowicz and M. Izydorek [IzM] in terms of the Euler class (see also [Mar3, Fo]). K. Komiya [Kom1] has suggested a necessary condition for the existence of an equivariant map between representation spheres using equivariant K-theory.

One can find the systematical study of the existence of equivariant maps between representation spheres in T. Bartsch's monograph [Bar1] (see also the corresponding bibliography therein).

5.6.5.   The simple proof of the Atiyah-Tall Theorem as well as Theorem 5.8 were suggested by A. Kushkuley.

References

[ADN]   Agmon, S., Douglis, A. and Nirenberg L., Estimates near the boundary
           for solutions of elliptic equations. *Comm. Pure and Appl. Math.* **2**
           (1959), 623-727.

[AP]   Arkhangelskii, A. and Ponomarev, V., *Fundamentals of general topol-
           ogy. Problems and exercises.* D. Reidel Publishing Company, Dor-
           drecht/Boston/ Lancaster 1984.

[AT]   Atiyah, M. and Tall, D., Group representations, λ-rings and the *J*-
           homomorphism. *Topology* **8** (1969), 253-297.

[Ba1]   Balanov, Z., *Geometric methods in a theory of equivariant vector fields.*
           Riga Polyt. Inst., Riga 1988 (in Russian).

[Ba2]   Balanov, Z., *Comparison principle for equivariant vector fields in non-
           linear analytic problems.* Thesis, Riga Polytechnic Institute, Riga 1989
           (in Russian).

[Ba3]   Balanov, Z., On the Krasnoselski Comparison Principle in infinite di-
           mensional spaces for the compact groups. In: *Latv. Matematičeshiy
           Yežegodnik*, Riga (1988), 207-215 (in Russian).

[Ba4]   Balanov, Z., On the finite coverings of an n-dimensional sphere and
           the Ljusternik Theorem. In: *Topologičeskie strukturi i ih otobraženya*,
           Latv. Gos. Univ., Riga (1987), 27-33 (in Russian).

[BA1]   Balanov, Z. and Ayevski, V., On the solutions of nonlinear elliptic equa-
           tions with symmetries. In: *XIV Škola po teorii operatoror v funkcional-
           nih prostranstvah*, Novgorodskii Pedagogičeskii Institut, Novgorod
           (1989), 7-8 (in Russian).

[BA2]   Balanov, Z. and Ayevski, V., On the solutions of nonlinear systems
           of elliptic equations with group symmetries. *Acta comment. Univ.
           Tartuensis* **960** (1993), 3-12.

[BB1]   Balanov, Z. and Brodsky, S., On the comparison principle of Krasnosel-
           skii. In: *VIII Škola po teorii operatorov v funkcionalnih prostranstvah*,
           Latv. Gos. Univ., Riga (1983), 15-16 (in Russian).

[BB2]   Balanov, Z. and Brodsky, S., Comparison principle of Krasnoselski and extension of equivariant maps. In: *Funkcionalniy Analiz. Teoriya Operatorov*, Ulyanovskii Pedagogičeskii Institut, Ulyanovsk (1984), 18-31 (in Russian).

[BB3]   Balanov, Z. and Brodsky, S., On the genus of some subsets of G-spheres. *Topol. Meth. Nonl. Anal.* **5** (1995), 101-110.

[BK1]   Balanov, Z. and Kushkuley, A., Comparison principle for equivariant maps into sphere. In: *Problemi čistoi i prikladnoi matematiki*, Tulskii Polytechn. Inst., Tula (1988), 28-33 (in Russian).

[BK2]   Balanov, Z. and Kushkuley, A., Comparison principle for equivariant maps. *Abstract of AMS* **17** (1988), 98.

[BK3]   Balanov, Z. and Kushkuley, A., On the problem of equivariant homotopoic classification. *Archiv der Mathematik* **65** (1995), 546-552.

[BKZ1]   Balanov, Z., Kushkuley, A. and Zabrejko, P., Geometric methods in a degree theory for equivariant maps. *Bochum, Preprint of Ruhr University, Heft Nr.* **137** (1990).

[BKZ2]   Balanov, Z., Kushkuley, A. and Zabrejko, P., Degree theory for equivariant maps: geometric approach. *Heidelberg, Topologie und nichtkommutative Geometrie, Heft Nr.* **47** (1992).

[BV1]   Balanov, Z. and Viničenko, S., A remark on comparison principle of Krasnoselskii for compact groups. In: *Topologičeskie prostranstva i ih otobraženya*, Latv. Gos. Univ., Riga (1985), 8-9 (in Russian).

[BV2]   Balanov, Z. and Viničenko, S., On the problem of the calculation of the winding number of an equivariant field. In: *Kačestvenniye i priblizenniye metodi issledovaniya operatornih uravnenii*, Yarosl. Gos. Univ., Yaroslavl (1985), 89-98 (in Russian).

[Bar1]   Bartsch, T., *Topological methods for variational problems with symmetries.* Lecture Notes in Mathematics 1560, Springer-Verlag, Berlin 1993.

[Bar2]   Bartsch, T., A simple proof of the degree formula for Z/p-equivariant maps. *Comment. Math. Helv.* **65** (1990), 85-95.

[Be]   Benci, V., A geometrical index for the group S^1 and some applications to the study of periodic solutions of ordinary equations. *Comm. Pure Appl. Math.* **34** (1981), 393-432.

[BoI]   Borisovich, J. and Izrailevich, J., Using spectral sequences for calcula-
             tion of a degree of an equivariant map. *Zapiski Voronezh. Univ.* **10**
             (1973), 1-12 (in Russian).

[BoISc]   Borisovich, J., Izrailevič, J. and Ščelokova, On the method of A. Borel
             spectral sequence in the theory of equivariant mappings. *Uspehi Mat.
             Nauk* **32** (1977), 161-162 (in Russian).

[BF]   Borisovich, J. and Fomenko, T., Homological methods in a theory of pe-
             riodic and equivariant mappings. In: *Global Analysis and Mathematical
             Physics,* Voronezh State Univ., Voronezh. (1987), 3-25 (in Russian).

[BZS]   Borisovich, J., Zvyagin, V. and Sapronov, J. Nonlinear Fredholm maps
             and the Leray-Schauder theory. *Uspehi Mat. Nauk* **32** (1977), 3-54 (in
             Russian).

[Bor1]   Borsuk, K., *Theory of Retracts.* Warszawa 1971.

[Bor2]   Borsuk, K., Uber Zerlegung einer euklidischen *n*-dimensionalen Vol-
             lkugeln in *n* Mengen. In: *Verhandlungen des Internationalen Mathe-
             matiker Kongresses, Zürich,* 1932; II. Band: *Sekt.-Vorträge, Orel Fus-
             sli,* Zürich, 1932, 142-198.

[Bor3]   Borsuk, K., Drei Sätze über die *n*-dimensionale euklidische Sphäre.
             Fund. Math. **20** (1933), 177-190.

[Bo1]   Bourbaki, N., *General Topology.* Part I, Hermann 1966.

[Bo2]   Bourbaki, N., *Groupes et Algebres de Lie.* Chap. 9, Masson 1982.

[Bow1]   Bowszyc, C., On the winding numbers and equivariant homotopy clas-
             ses of maps of manifolds with some finite group actions. *Fund. Math.*
             115 (1983), 235-247.

[Bow2]   Bowszyc, C., Differential methods in equivariant and non-equivariant
             topology. In: *Topics in equivariant topology,* Sem. Math. Sup. **108**,
             Montreal (1989), 57-104.

[Bre]   Bredon, G.E., *Introduction to compact transformation groups.* Aca-
             demic Press 1972.

[BtD]   Bröcker, T. and tom Dieck, T., *Representations of compact Lie groups.*
             Springer, New York 1985.

[Dac]   Daccach, J., Nonexistence of equivariant degree one maps. *Proc. of
             AMS* **101** (1987), 530-532.

[Di1]   tom Dieck, T., *Transformation Groups and Representation Theory*.
             Lecture Notes in Mathematics 766, Springer Verlag, Berlin 1979.

[Di2]   tom Dieck, T., *Transformation Groups*. de Gruyter, Berlin 1987.

[Do1]   Dold, A., *Lectures on algebraic topology*. Springer Verlag, Berlin 1972.

[Do2] .   Dold, A., Simple proofs of Borsuk-Ulam results. *Contemp. Math.* **19**
             (1983), 65-69.

[Dan]   Dancer, E., Symmetries, degree, homotopy indices and asymptotically
             homogeneous problems. *Nonl. Anal., Theory, Meth. and Appl.* **6**
             (1982), 667-686.

[DFN]   Dubrovin, B., Fomenko, A. and Novikov, S., *Modern Geometry – Meth-
             ods and Applications, Part II. The Geometry and Topology of Mani-
             folds*. Springer Verlag, Berlin-New York 1984.

[Ei]   Eilenberg, S., On the theorem of P.A. Smith concerning fixed points
             for periodic maps. *Duke Mathematical Journal* **6** (1940), 428-437.

[En]   Engelking, R., *General topology*, Warszawa 1977.

[Fa1]   Fadell, E., The equivariant Ljusternik-Schnirelman methods for invari-
             ant functionals and relative cohomological index theories. In: *Méth.
             topologiques en analyse non linéaires*, A. Granas (ed.), Sémin. Math.
             Sup., No **95** Montréal 1985, 41-70.

[Fa2]   Fadell, E., The relationship between Ljusternik-Schnirelman category
             and the concept of genus. *Pacific J. Math.* **89** (1980), 283-319.

[Fa3]   Fadell, E., Cohomological methods in non-free G-spaces with applica-
             tions to general Borsuk-Ulam theorems and critical point theorems for
             invariant functionals. In: *Nonlinear Functional Analysis and its Appli-
             cations*, S. P. Singh (ed.), Proc. Maratea 1985, NATO ASI Ser. **C 173**
             Reidel, Dordrecht 1986, 1-45.

[FH1]   Fadell, E. and Husseini, S., Relative cohomological index theories. *Adv.
             in Mathematics* **64** (1987), 1-31.

[FH2]   Fadell, E. and Husseini, S., An ideal valued cohomological index theory
             with applications to Borsuk-Ulam and Bourgain-Yang theorems. *Ergod.
             Th. and Dynam. Sys.* 8 (1988), 73-85.

[FR]   Fadell, E., Rabinowitz, P., Generalized cohomological index theories
             for Lie group actions with an application to bifurcation questions for
             Hamiltonian systems. *Invent. Math.* **45** (1978), 139-174.

[FHR]   Fadell, E., Husseini, S. and Rabinowitz, P., Borsuk-Ulam theorems for
             arbitrary S^1-action and applications. *Trans. Amer. Math. Soc* **274**
             (1982), 345-360.

[Fe]   Fet, A., Involutonary mappings and coverings of sphere. In: *Trudy
             Sem. Funct. Anal., Voronez Gos. Univ.* **1** (1956), 536-552.

FPR   Fitzpatrick. P.M., Pejsachowicz. J. and Rabier, P.J., The degree of
             proper C^2 Fredholm maps: covariant theory. *Topol. Meth. Nonl.
             Anal.* **3** (1993), 325-367.

[Fo]   Fomenko, T., Algebraic properties of certain cohomological invariants
             of equivariant maps. *Matemat. zametki* **50** (1981), 108-117 (in Rus-
             sian).

[Ge]   Gelfand, I., *Lectures on linear algebra.* Interscience Publ., New York,
             1961.

[GKW]   Gęba, K., Krawcewicz, W. and Wu, J., An equivariant degree with
             applications to symmetric bifurcation problems. Part 1: Construction
             of the degree *Bull. London Math. Soc.,* **69** (1992), 377-398.

[Got]   Gottlieb, D., The trace of an action and the degree of a map. *Trans.
             Amer. Math. Soc.* **287** (1985), 419-429.

[Hör]   Hörmander, L., *Linear partial differential operators.* Springer, Berlin
             1963.

[Hu]   Hu, S.T., *Homotopy theory.* Academic Press, New York 1959.

[Hus]   Husemoller, D., *Fiber bundles.* Springer, Berlin 1974.

[IV]   Ize, J. and Vignoli, A., Equivariant degree for abelian actions. Part I:
             equivariant homotopy groups. *TMNA,* **2** (1993), 367-413.

IV1]   Ize, J. and Vignoli, A., Preprint.

[IMV1]   Ize, J., Massabó, I. and Vignoli, A., *Degree theory for equivariant maps.
             I.* Trans. Amer. Math. Soc. **315** (1992), 433-510.

[IMV2]   Ize, J., Massabó, I. and Vignoli, A., *Degree theory for equivariant maps,
             the general S^1-actions.* Memoirs AMS **481** (1992).

[Iz]   Izrailevič, K., The index of a semi-free periodic mapping. *Matemat.
             Zametki* **13** (1973), 46-50.

[IM]   Izrailevič, J. and Muhamadiev, E., On the theory of periodic mappings
             of spheres In: *Seventh Mathematical Summer School (Kaciveli, 1969).*

Izdanie Mat. Akad. Nauk Ukrain SSR, Kiev, 1970, 295-305 (in Russian).

[IO1]   Izrailevič, J. and Obuhovskii, V., On equivariant multivalued mappings *Soviet Math. Dokl.* **13** (1972), 864-867.

[IO2]   Izrailevič, J. and Obuhovskii, V., Certain topological characteristics of equivariant multivalued mappings. In: *Trudy Mat. Fak. Voronez. Gos. Univ.* **10** (1973), 52-61 (in Russian).

[IzM]   Izydorek, M. and Marzantowitz, W., Equivariant maps between cohomology spheres. *Heidelberg, Forschungsschwerpunkt Geometrie, Heft Nr.* **11** (1990).

[Ja1]   Jaworowski, J., Extensions of *G*-maps and Euclidean *G*-retracts. *Math. Z.* **146** (1976), 143-148.

[Ja2]   Jaworowski, J., An equivariant extension theorem and *G*-retracts with a finite structure *Manuscripta Math.* **35** (1981), 323-329.

[Ka]   Kato, T., *Perturbation theory for linear operators.* Springer, Berlin 1966.

[Kom]   Komiya, K., Fixed point indices of equivariant maps and Möbius inversion *Inv. Math.* **91** (1988), 129-135.

[Kom1] Komiya, K., Equivariant *K*-Theory and Maps between Representation Sphe- res. Preprint (1994).

[Kos]   Kostadinov, S., On a special case of comparison of equivariant vector field on a finite-dimensional sphere. *Zapiski Plovdiv Univ.* **22** (1984), 79-86 (in Russian).

[Kr1]   Kranoselskii, M., On computation of the rotation of a vector field on the *n*-dimensional sphere. *Dokl. Akad. Nauk SSSR* **101** (1955), 401-404 (in Russian).

[Kr2]   Krasnoselskii, M., *Topological Methods in the Theory of Non-linear Integral Equations.* Pergamon Press, Oxford-London-New York-Paris 1964.

[Kr3]   Krasnoselskii, M., On special coverings of a finite-dimensional sphere. *Dokl. Akad. Nauk SSSR* **103** (1955), 961-964 (in Russian).

[KZ]   Krasnoselskii, M. and Zabrejko, P., *Geometrical Methods of Nonlinear Analysis.* Springer, Berlin 1984.

[KB]   Kushkuley, A. and Balanov, Z., A comparison principle and extension
            of equivariant maps. *Manusscr. Math.* **83** (1994). 239-264.

[Ku]   Kurosch, A., *The theory of groups.* V.I, New York. Chelsea 1955.

[Lai]   Laitinen, E., Unstable homotopy theory of homotopy representations.
            In: Lecture Notes in Math. 1217, S. Jackowski and Pawalowski (ed.),
            Springer, Berlin (1986), 210-248.

[La]   Lashof, R., The equivariant extension theorem. *Proc. Amer. Mat.
            Soc.* **83** (1981), 138-140.

[LW]   Lee, C.N. and Wasserman, A.. *On the groups JO(G).* Mem. Amer.
            Mat. Soc. 159, Providence, Rhode Island 1975.

[Le]   Lefschetz, S., *Algebraic topology.* Amer. Math. Soc., New York 1942.

[Lü]   Lück, W., The equivariant degree. In: *Algebraic topology and transfor-
            mation groups, Proc., Göttingen 1987*, Lecture. Notes in Math. 1361,
            123-166.

[LS]   Lusternik, L. and Schnirelman, A., *Métodes Topologiques dans des
            Problèmes Variationels.* Hermann, 1934.

[Mc]   Mc-Crory, C., Stratified general position. In: *Algebraic and Geometric
            Topology*, Lecture Notes in Mathematics 664, Springer, Berlin 1978.

[Mad1]   Madirimov, M., On the Jaworowski Theorem on the extension of equiv-
            ariant mappings. *Uspehi Mat. Nauk* **39** (1984), 147-148 (in Russian).

[Mad2]   Madirimov, M., *Dimension and retraction in a theory of the topological
            transformation groups.* Fan, Tashkent 1986 (in Russian).

[Man]   Mann, L., Finite orbit structure on locally compact manifolds. *Trans.
            Amer. Math. Soc.* **99** (1961), 41-59.

[Mar1]   Marzantowicz, W., On the nonlinear elliptic equations with symmetry.
            J. Math. Anal. Appl. 81 (1981), 156-181.

[Mar2]   Marzantowicz, W., A G-Lusternik-Schnirelman category of a space with
            an action of a compact Lie group. *Topology* **28** (1989), 403-412.

[Mar3]   Marzantowicz, W., Borsuk-Ulam theorem for any compact Lie group.
            J. London Math. Soc. **49** (1994), 195-208.

[MW]   Mawhin, J. and Wilem, M., *Critical Point Theory and Hamiltonian
            Systems.* Springer, New York 1989.

[Mo]   Mostov, G., Equivariant embeddings in Euclidean space. *Ann. of Math.*
            65 (1957), 432-446.

[Mon]   Montgomery, D., Orbits of highest dimensions. In: *Seminar on trans-*
            formation groups, Princeton Univ. Press, New York (1960), 79-97.

[Mor]   Morita, K., On the dimension of normal spaces I. *Journ. Japan Math.*
            20 (1950), 1-36.

[Mu]   Muhamadiev, E., On a theory of periodic completely continuous vector
            fields. *Uspehi Mat. Nauk* **22** (1967), 127-128 (in Russian).

[Ni1]   Nirenberg, L.. *Topics in nonlinear functional analysis.* New York Uni-
            versity, Courant Inst. of Math. Sc., New York 1974.

[Ni2]   Nirenberg, L., Comments on nonlinear problems. *Mathematische (Ca-*
            tania) **36** (1981), 109-119.

[Po]   Pontrjagin, L., *Topological groups.* Princeton University Press, Prince-
            ton 1946.

[Rab1]   Rabier, P., Topological degree and the theorem of Borsuk for general
            covariant mappings with applications. *Nonlinear anal.* **16** (1991), 399-
            420.

[Rab2]   Rabier, P., Symmetries, topological degree and a theorem of Z. Q.
            Wang. (to appear in "Rocky Montain J. Math.")

[Ra1]   Rabinowitz, P., A note on a nonlinear elliptic equations. *Indiana Univ.*
            Math. J. **22** (1972), 43-49.

[Ra2]   Rabinowitz, P., *Minimax methods in critical point theory with applica-*
            tion to differential equations. CBMS, Regional Confer. Ser. in Math.
            65, Amer. Math. Soc.. Providence, R.I.. 1986.

[Ru]   Rubinsztein, R., *On the equivariant homotopy of spheres.* Dissertation,
            Dissert. Math. 134, Poland 1976.

[Sc1]   Schelokova, T., Floyd-Smith theory and equivariant maps of manifolds.
            In: *Zapiski Voronezhskogo Univ.* (1974), 57-69. (in Russian).

[Sc2]   Schelokova, T., Towards a theory of equivariant maps of cohomological
            spheres. In: *Operator equations,* Voronezh. Univ. (1978), 36-64 (in
            Russian).

[Sc3]   Schelokova, T., On calculation of degrees of Z_p-equivariant maps. *Za-*
            piski NIIM, Voronezh University **20** (1978), 38-52 (in Russian).

[Sc4]   Schelokova, T., On a problem of calculation of a degree of an equivariant
          map. *Sib. Mat. Žurnal* **19** (1978), 38-52 (in Russian).

[Se]   Serre, J.-P., *Linear representations of finite groups.* Springer, Berlin
          1977.

[Sn]   Snaith, V., *J*-equivalence of group representations. *Proc. Camb. Phil.
          Soc.* **70** (1971), 9 - 14.

[Sp]   Spanier, E., *Algebriac topology.* Springer, New York 1989.

[Šv]   Švartz, A., A genus of a fiber bundle space. *Trudy Mosc. Mat. Obs.*
          10 (1961), 217-272 , and **11** (1962), 99-126 (in Russian); English trans-
          lation in *Transl. Amer. Math. Soc., II. Ser.,* **55** (1966), 49-140.

[St]   Steinlein, H., Borsuk's antipodal theorem and its generalizations and
          applications: a survey. In: *Méth. topologiques en analyse nonlinéaires,*
          A. Granas (ed.), Sémin. Math. Sup. No **95** Montréal 1985, 166-235.

[Str]   Struwe, M., *Variational Methods.* Springer, Berlin 1990.

[To]   Tornehave, J., Equivariant maps of spheres with conjugate orthogonal
          actions. In: *Algebraic Topology, Proc. Conf. London Ont. 1981,
          Canadian Math. Soc. Conf. Proc.* **2**, part 2 (1982), 275-301.

[Wan]   Wang, Z. Q., Symmetries and calculation of the degree. *Chinese Ann.
          Math.* **10B** (1989), 520-536.

[We]   Wei-Yue Ding, Generalizations of the Borsuk Theorem. *J. of Mat.
          Anal. and Appl.* **110** (1985), 553-567.

[Za1]   Zabrejko, P., Towards a theory of periodic vector fields. *Vestn. Jaros-
          lavl. Univ.* **2** (1973), 23-30 (in Russian).

[Za2]   Zabrejko, P., Towards the homotopy theory of periodic vector fields.
          In: *Geometric methods in algebraic and analytic problems,* Jaroslavl
          (1980), 116-119.

[ZK]   Zabrejko, P. and Kostadinov, S., On winding numbers of equivariant
          vector fields. *Zapiski Belorussk. Univ.* **3** (1986), 57-60 (in Russian).

[Zee]   Zeeman, E., *Seminar on Combinatorial Topology.* Institut des Hautes
          Etudes Scientifiques, Paris 1963.

Subject index

Lecture Notes in Mathematics

For information about Vols. 1–1449
please contact your bookseller or Springer-Verlag

Vol. 1491: E. Lluis-Puebla, J.-L. Loday, H. Gillet, C. Soulé, V. Snaith, Higher Algebraic K-Theory: an overview. IX, 164 pages. 1992.

Vol. 1492: K. R. Wicks, Fractals and Hyperspaces. VIII, 168 pages. 1991.

Vol. 1493: E. Benoît (Ed.), Dynamic Bifurcations. Proceedings, Luminy 1990. VII, 219 pages. 1991.

Vol. 1494: M.-T. Cheng, X.-W. Zhou, D.-G. Deng (Eds.), Harmonic Analysis. Proceedings, 1988. IX, 226 pages. 1991.

Vol. 1495: J. M. Bony, G. Grubb, L. Hörmander, H. Komatsu, J. Sjöstrand, Microlocal Analysis and Applications. Montecatini Terme, 1989. Editors: L. Cattabriga, L. Rodino. VII, 349 pages. 1991.

Vol. 1496: C. Foias, B. Francis, J. W. Helton, H. Kwakernaak, J. B. Pearson, H_∞-Control Theory. Como, 1990. Editors: E. Mosca, L. Pandolfi. VII, 336 pages. 1991.

Vol. 1497: G. T. Herman, A. K. Louis, F. Natterer (Eds.), Mathematical Methods in Tomography. Proceedings 1990. X, 268 pages. 1991.

Vol. 1498: R. Lang, Spectral Theory of Random Schrödinger Operators. X, 125 pages. 1991.

Vol. 1499: K. Taira, Boundary Value Problems and Markov Processes. IX, 132 pages. 1991.

Vol. 1500: J.-P. Serre, Lie Algebras and Lie Groups. VII, 168 pages. 1992.

Vol. 1501: A. De Masi, E. Presutti, Mathematical Methods for Hydrodynamic Limits. IX, 196 pages. 1991.

Vol. 1502: C. Simpson, Asymptotic Behavior of Monodromy. V, 139 pages. 1991.

Vol. 1503: S. Shokranian, The Selberg-Arthur Trace Formula (Lectures by J. Arthur). VII, 97 pages. 1991.

Vol. 1504: J. Cheeger, M. Gromov, C. Okonek, P. Pansu. Geometric Topology: Recent Developments. Editors: P. de Bartolomeis, F. Tricerri. VII, 197 pages. 1991.

Vol. 1505: K. Kajitani, T. Nishitani, The Hyperbolic Cauchy Problem. VII, 168 pages. 1991.

Vol. 1506: A. Buium, Differential Algebraic Groups of Finite Dimension. XV, 145 pages. 1992.

Vol. 1507: K. Hulek, T. Peternell, M. Schneider, F.-O. Schreyer (Eds.), Complex Algebraic Varieties. Proceedings, 1990. VII, 179 pages. 1992.

Vol. 1508: M. Vuorinen (Ed.), Quasiconformal Space Mappings. A Collection of Surveys 1960-1990. IX, 148 pages. 1992.

Vol. 1509: J. Aguadé, M. Castellet, F. R. Cohen (Eds.), Algebraic Topology - Homotopy and Group Cohomology. Proceedings, 1990. X, 330 pages. 1992.

Vol. 1510: P. P. Kulish (Ed.), Quantum Groups. Proceedings, 1990. XII, 398 pages. 1992.

Vol. 1511: B. S. Yadav, D. Singh (Eds.), Functional Analysis and Operator Theory. Proceedings, 1990. VIII, 223 pages. 1992.

Vol. 1512: L. M. Adleman, M.-D. A. Huang, Primality Testing and Abelian Varieties Over Finite Fields. VII, 142 pages. 1992.

Vol. 1513: L. S. Block, W. A. Coppel, Dynamics in One Dimension. VIII, 249 pages. 1992.

Vol. 1514: U. Krengel, K. Richter, V. Warstat (Eds.), Ergodic Theory and Related Topics III. Proceedings, 1990. VIII, 236 pages. 1992.

Vol. 1515: E. Ballico, F. Catanese, C. Ciliberto (Eds.), Classification of Irregular Varieties. Proceedings, 1990. VII, 149 pages. 1992.

Vol. 1516: R. A. Lorentz, Multivariate Birkhoff Interpolation. IX, 192 pages. 1992.

Vol. 1517: K. Keimel, W. Roth, Ordered Cones and Approximation. VI, 134 pages. 1992.

Vol. 1518: H. Stichtenoth, M. A. Tsfasman (Eds.), Coding Theory and Algebraic Geometry. Proceedings, 1991. VIII, 223 pages. 1992.

Vol. 1519: M. W. Short, The Primitive Soluble Permutation Groups of Degree less than 256. IX, 145 pages. 1992.

Vol. 1520: Yu. G. Borisovich, Yu. E. Gliklikh (Eds.), Global Analysis – Studies and Applications V. VII, 284 pages. 1992.

Vol. 1521: S. Busenberg, B. Forte, H. K. Kuiken, Mathematical Modelling of Industrial Process. Bari, 1990. Editors: V. Capasso, A. Fasano. VII, 162 pages. 1992.

Vol. 1522: J.-M. Delort, F. B. I. Transformation. VII, 101 pages. 1992.

Vol. 1523: W. Xue, Rings with Morita Duality. X, 168 pages. 1992.

Vol. 1524: M. Coste, L. Mahé, M.-F. Roy (Eds.), Real Algebraic Geometry. Proceedings, 1991. VIII, 418 pages. 1992.

Vol. 1525: C. Casacuberta, M. Castellet (Eds.), Mathematical Research Today and Tomorrow. VII, 112 pages. 1992.

Vol. 1526: J. Azéma, P. A. Meyer, M. Yor (Eds.), Séminaire de Probabilités XXVI. X, 633 pages. 1992.

Vol. 1527: M. I. Freidlin, J.-F. Le Gall, Ecole d'Eté de Probabilités de Saint-Flour XX – 1990. Editor: P. L. Hennequin. VIII, 244 pages. 1992.

Vol. 1528: G. Isac, Complementarity Problems. VI, 297 pages. 1992.

Vol. 1529: J. van Neerven, The Adjoint of a Semigroup of Linear Operators. X, 195 pages. 1992.

Vol. 1530: J. G. Heywood, K. Masuda, R. Rautmann, S. A. Solonnikov (Eds.), The Navier-Stokes Equations II – Theory and Numerical Methods. IX, 322 pages. 1992.

Vol. 1531: M. Stoer, Design of Survivable Networks. IV, 206 pages. 1992.

Vol. 1532: J. F. Colombeau, Multiplication of Distributions. X, 184 pages. 1992.

Vol. 1533: P. Jipsen, H. Rose, Varieties of Lattices. X, 162 pages. 1992.

Vol. 1534: C. Greither, Cyclic Galois Extensions of Commutative Rings. X, 145 pages. 1992.

Vol. 1535: A. B. Evans, Orthomorphism Graphs of Groups. VIII, 114 pages. 1992.

Vol. 1536: M. K. Kwong, A. Zettl, Norm Inequalities for Derivatives and Differences. VII, 150 pages. 1992.

Vol. 1537: P. Fitzpatrick, M. Martelli, J. Mawhin, R. Nussbaum, Topological Methods for Ordinary Differential Equations. Montecatini Terme, 1991. Editors: M. Furi, P. Zecca. VII, 218 pages. 1993.

Vol. 1538: P.-A. Meyer, Quantum Probability for Probabilists. X, 287 pages. 1993.

Vol. 1539: M. Coornaert, A. Papadopoulos, Symbolic Dynamics and Hyperbolic Groups. VIII, 138 pages. 1993.